PRINCIPLES OF
BIOLOGICAL REGULATION

An Introduction to Feedback Systems

Richard W. Jones

Electrical Engineering Technological Institute
Northwestern University
Evanston, Illinois

ACADEMIC PRESS New York and London   1973

ACADEMIC PRESS, INC.
111 Fifth Avenue, New York, New York 10003

United Kingdom Edition published by
ACADEMIC PRESS, INC. (LONDON) LTD.
24/28 Oval Road, London NW1

LIBRARY OF CONGRESS CATALOG CARD NUMBER: 72-12283

PRINTED IN THE UNITED STATES OF AMERICA

Contents

1. Regulatory Processes in Biological Systems

2. Flow Processes in the Steady State

3. Dynamic Behavior; the Transient Response

4. Introduction to Feedback; the Steady State

5. Feedback Systems; Dynamic Behavior

6. Sinusoidal Signals

7. Stability

8.   Distinctive Features of Homeostatic Systems

9.   Nonlinear Systems

10.   Biochemical Control

Preface

The objective of this book is to provide the life scientist with some understanding of control, regulatory, and feedback mechanisms in biological systems. The concepts to be introduced are related to the dynamic behavior of both individual biological processes and systems of processes that make up an organism. A further objective is the description of characteristics of biological feedback systems with emphasis upon the physical concepts, and thus with a minimum of mathematical formalism. It will become clear, however, that although a qualitative treatment of these ideas is adequate for some discussions, it is not sufficient for those concepts that turn upon the relative magnitudes of several quantities. As a consequence, it is hoped that the reader will sense the need for mathematical tools and techniques of greater power if he is to refine his understanding of the behavior of complex systems having many component processes.

The terms biological and physiological are used somewhat interchangeably. Although most of the examples are physiological, many of the concepts can be extended to a variety of biological problems, as is suggested on occasion. Genetics, ontogeny, and ecology are all fields in which many of the concepts described have already appeared.

At many points the discerning reader may detect exceptions to the relations and concepts described. With the sole view of presenting basic concepts in a simplified language, I have taken the liberty of making rather general statements on occasion without surrounding them with adequate restrictions and qualifying phrases. The hazards are obvious, but at the same time the reader is hereby warned that constraints on the general applicability of many statements do exist, and he is urged to seek out the exceptions or limitations. And perhaps he may even be encouraged to delve deeper into many topics so as to confront the author at some future encounter. As an aid in such further study a brief résumé of the salient points in a mathematical analysis will be found at the end of most chapters.

With the above objectives, it has been impossible to describe specific physiological examples in great detail. To do so would have expanded this text beyond all reasonable bounds. It is probably accurate to state that no one of the physiological systems has been described with the completeness and detail needed for clinical use, but it is hoped that this volume may contribute in some small way to that development.

Physiological examples have been introduced at numerous points in the hope that they will lend substance to the argument. For the most part they have been chosen to illustrate some system concept, however crudely. In most instances the examples have been greatly simplified, but not, it is hoped, to the point of unrecognizability or irrelevance to life processes.

Acknowledgments

This book is dedicated to the reader in the hope that it may smooth his passage through a portion of the vast literature of physiology, guided by the one concept of homeostasis. However, this dedication would be an impossibility were it not for the continued guidance of my mentors, especially Dr. John S. Gray, Dr. Christina Enroth-Cugell, and Dr. Miriam Eubank-Jones. To this I can only add my thanks to many classes of students who have patiently watched this text grow from small beginnings, and unknowingly contributed to that development.

Regulatory Processes in Biological Systems | *Chapter 1*

Biology as a Science of Organization   1.1

Current developments in the biological sciences make it increasingly clear that organization, or system theory, is a fundamental component of biology. It is axiomatic that biology is rooted in physical and chemical processes, but classical physics and chemistry are not enough to describe the evolution, development, and physiology of an organism. The component processes, such as diffusion, mass transport, and protein synthesis, are inadequate in themselves to account for the physiology of the entire organism, to say nothing about its organized behavior. To these components must be added the organizational framework—the manner in which the subsidiary processes are related and coupled. Numerous writers have described the two extreme viewpoints, that of molecular biology on the one hand, and organismic, systemic biology on the other. Simpson (1963, 1967) has provided an especially thoughtful examination of this situation, and has even elevated it to the position of a crisis in biology.

There are several reasons for this seeming concern. In the first place, an understanding of biological organization is only now developing, and so a full description of the role of organization in biological processes is at most in embryonic form. Second, in systems of many components, organization connotes complexity, and there is some feeling that we do not have the language necessary to describe these highly interrelated processes. At best, the mathematical language which appears most suitable at the present time is not an intellectual tool of most life scientists.

Among the many types of organization that characterize biological processes in cell, organism, and population, that termed regulatory, or homeostatis, is most ubiquitous. This fact, together with some similarity to existing technological systems, makes regulatory biology a reasonable first choice in developing a broader theory of biological organization.

There are four factors that prompted the writing of this book.

(1) It is apparent that feedback system theory is becoming of increasing significance to most life scientists, whether they are biologists concerned with the classical problems of cell, organ, organism, and population, or medical practioners concerned with human problems of health and disease. The wide prevalence of regulatory processes, and in particular the feedback mechanisms with which they are associated, make a study of such systems mandatory. It is possibly the first item on any agenda devoted to a study of organization.

(2) With very few exceptions, existing texts develop the concepts of regulatory behavior with the techniques, terminology, and mathematics of the engineering profession. The present text has been written with the hypothesis that most of the concepts can be developed in a qualitative nonmathematical form that is intuitively clear. The emphasis, then, is upon physical concepts, presented in such a manner (it is hoped) that the basic ideas can be translated into a biological context. How-

ever, biological systems do not lend themselves to a direct application of engineering concepts, but rather require careful study to assure that the result has biological relevance. It may well be that significant modification and extension of basic feedback theory will evolve as a consequence.

(3) The existence of feedback in any physical system has profound effects on its behavior, so that the responses to stimuli and disturbances are quite unlike those of individual components. Feedback accentuates certain properties of a system and endows it with others it did not have. Furthermore, feedback greatly complicates the interpretation of experimental findings, since a given response may not be attributable solely to the component being measured because it may also contain contributions from components located remotely in the feedback loop. These aspects of feedback suggest that experimental biologists need at least a basic understanding of feedback phenomena.

(4) The concept of regulation or homeostasis has been extended in the biological literature far beyond its original context. While some of these extensions offer exciting vistas to the investigator, others seem to cloud the issue and lead to an imprecision of discourse that impedes the development of clear ideas. In this and later chapters, some definitions will be discussed with a view to clarifying the discussion and promoting further study of regulatory problems and their bearing upon organization.

Although these factors appear to me as sufficient justification for another book, it should also be pointed out that the framework of this text, as well as the present state of knowledge, imposes severe limitations. Our knowledge of homeostasis is still extremely elementary and the reader is here forewarned that feedback theory does not provide the one ingredient that will unlock (or unleash) all biological secrets. Rather, it should be looked upon as another tool to assist in finding some answers, to lead on occasion to new and intriguing questions, and to provide integrating viewpoints within the family of biological sciences. An understanding of dynamic behavior and of feedback theory is but another tool for the life scientist. It is a tool in the same sense as is a microscope that reveals structure, the spectrograph as it depicts composition, or the recording electrode measuring neural signals. Each of these methods permits the examination of one facet of life, and each contributes to a deeper understanding of the whole.

Regulatory Biology 1.2

Although more precise definitions will be found in subsequent chapters, the following terms are introduced here in a descriptive manner. *Regulation,* when used to describe the property of a *system,* refers to the fact that certain changes brought about by disturbances to the system are minimized. Regulation may also refer to the *mechanisms* by which this minimization is effected. The terms *homeostasis* and *homeostat* are used in similar senses, but in a biological context. There is a relative component to these terms that implies the possibility of specifying those portions of the system that endow it with this property of regulation. To say that a given system is regulated is meaningless, unless a mechanism is identified in whose absence the system becomes less well regulated. In a broad sense, the homeostat has acquired mechanisms that improve its ability to resist disturbances.

Homeostatic mechanisms are the means whereby organisms have achieved relative freedom from the constraints of their environments. Having acquired greater freedom in this manner, that is, by evolving homeostatic processes, the organism has acquired a greater variety of functions, not only those required to maintain the homeostasis, but also those permitted by the enlarged environmental exposure.

This independence of the environment is a relative matter. Organisms do not become completely independent, but only more nearly independent of environmental factors, as they acquire the ability to maintain a constant internal environment in which most of their cells live. One finds that organisms have acquired not one homeostat, but a very large number, each related to some physical aspect of their internal affairs or of their behavior regarding the environment. Furthermore, each of the homeostats has its own limits within which it can maintain its integrity, and excursions beyond these limits are almost invariably fatal.

The biological quantities that are regulated fall into three broad categories, and we will want to discuss not only the differences but the similarities in the regulatory processes for these several types of regulating systems. In addition, we shall pay attention to the aspects which distinguish biological regulators from their technological counterparts. Although extended use will be made of analogous processes, the reader is forewarned that the use of the mistaken analog is a grave and present danger to biology.

The *quantity or concentration of the materials* making up the organism, that is, the many types of molecules or chemical species, is for the most part closely regulated by means of a large number of homeostatic processes. The need for regulation arises from the "mismatch" between supply and demand, and reflects changes in the intake, use, and excretion of specific molecules. Although the body contains stores of many chemical species, these cannot act as effective regulators without the addition of facultative processes to enhance or inhibit the synthesis and movement of material through appropriate pathways within the organism.

The *energy balance* within the body is carefully regulated in those organisms known as homeotherms. Changes in body temperature serve to indicate corresponding changes in the energy balance and thus in energy production, energy absorption, or energy dissipation. As is well known, slight changes in body temperature may be lethal, so that a means must be available for correcting any changes in the energy flow pattern in order not to upset the body temperature.

The various *muscular control mechanisms* involved in the control of posture and movement are also found to be well regulated, and thus we can subsume these processes under the term homeostasis.

An organism may exhibit the properties of homeostasis by means of a number of different mechanisms. Although the discussion of the first chapters will be related to feedback systems, it is well to point out that this is only one of several possible ways of controlling the perturbations in a biological system. Some of these other methods, such as rigidity, compensation, buffering, and stochastic regulation, are briefly discussed later.

The major portion of this text is devoted to homeostatic systems that function to maintain a relative constancy in a physical or chemical process. However, there is a growing awareness that this may be too limited a view. It is becoming increasingly apparent that oscillatory phenomena play a significant, if not an essential, role in many life processes. We refer here not to the obvious cardiac and respiratory rhythms, but to the so-called circadian (approximately 24-hr) rhythms which have been shown to exist widely in metabolic and activity patterns (Pittendrigh, 1961). The prevalence of this phenomenon has led to the hypothesis that organisms should be viewed as systems of coupled oscillators. This matter will be touched upon briefly in later chapters.

Historical Development 1.3

Like many fundamental concepts, biological regulation, or homeostasis, has an ancient and honorable ancestry. Its development from Greek to modern times has been admirably described by Adolph (1961), who has also discussed the many interpretations of the homeostatic concept and related ideas.

Possibly the first to recognize the prevalence of regulatory processes within the organism and to state the concept in reasonably precise and modern terms was Claude Bernard (1865, 1878).

The organism is merely a living machine so constructed that, on the one hand, the outer environment is in free communication with the inner organic environment, and, on the other hand, the organic units have protective functions, to place in reserve the materials of life and uninterruptedly to maintain the humidity, the warmth and other conditions essential to vital activity. Sickness and death are merely a dislocation or disturbance of the mechanism which regulates the contact of vital stimulants with organic units.

Bernard coined the phrase *fixite du milieu interieur* and stated that it was the condition for a free life.

The next major step in the development of this concept was taken by Cannon (1929), who coined the word *homeostasis,* which he defined as the "coordinated physiological reactions which maintain most of the steady states in the body." The term may be somewhat misleading, since the first portion *homeo* is an abbreviation meaning "like" or "similar, but not the same," and the second word *stasis* implies something static, immobile, and stagnant. It was not Cannon's intention to imply something static, and he points this out in his article. Cannon was well aware that these homeostatic mechanisms are dynamic in character, continually modifying various processes in order to meet the changing needs of the organism (Cannon, 1939).

Among the first attempts to refine these notions are the paper by Rosenblueth *et al.* (1943), and the book by Wiener (1948) that introduced the term *cybernetics.* Unfortunately, this term has been seized upon by many writers in as many fields, with the result that its precise meaning is now well-diluted and its utility is of doubtful value. The texts by Grodins (1963), Hughes (1964), Yamamoto and Brobeck (1965), Kalmus (1966), and Bayliss (1966) constitute significant contributions from biologists and physiologists to a growing borderland between biological and technological regulation. The texts by Milhorn (1966) and Milsum (1966) approach this territory with somewhat more of an engineering stance. The origins of selected homeostats are discussed by Adolph (1968).

Epitome 1.4

1.   An organism is viewed as an organized collection of many physico-chemical processes. The viability and behavior exhibited by the organism is a direct consequence of the manner in which these component processes are organized and interrelated.

2.   Among the many kinds of biological organization, those termed regulatory, or homeostatic, are prevalent at all biological levels. The similarities between these homeostatic systems and technological regulators provide a starting point for the study of homeostatic mechanisms.

3.   Biological regulation refers to those systems that can act to minimize the effects of disturbing environmental factors. A homeostatic system is one in which certain processes can be identified whose principal function is that of regulation.

4.   Homeostatic systems make extensive use of feedback as a control mechanism to effect regulation. Feedback systems may serve to regulate the concentration of chemical species, the energy level, or the position in neuromuscular control.

5.   A feedback system has properties that are greatly different from those of the individual components. This fact leads to advantages for the organism, but also to difficulties for the experimental biologist.

References

Adolph, E. F. (1961). Early concepts of physiological regulations. *Physiol. Rev.* **41**, 737–770.

Adolph, E. F. (1968). "Origins of Physiological Regulations." Academic Press, New York.

Bayliss, L. E. (1966). "Living Control Systems." English Univ. Press, London.

Bernard, C. (1865). "Introduction to the Study of Experimental Medicine." Reprinted, Dover, New York. 1957.

Bernard, C. (1878). *Les Phenomenes de la Vie,* **1,** 113, 121.

Cannon, W. B. (1929). Organization for physiological homeostasis. *Physiol. Rev.* **9**, 399–431.

Cannon, W. B. (1939). "The Wisdom of the Body." Norton, New York.

Grodins, F. S. (1963). "Control Theory and Biological Systems." Columbia Univ. Press, New York.

Hughes, G. M. (ed.) (1964). Homeostasis and Feedback Mechanisms. *Symp. Soc. Exp. Biol.* No. XVIII, Cambridge Univ. Press, London and New York.

Kalmus, H. (1966). "Regulation and Control in Living Systems." Wiley, New York.

Milhorn, H. T., Jr. (1966). "Application of Control Theory to Physiological Systems." Saunders, Philadelphia, Pennsylvania.

Milsum, J. H. (1966). "Biological Control System Analysis." McGraw-Hill, New York.

Pittendrigh, C. S. (1961). On temporal organization in living systems. *Harvey Lect.* **56**, 93–125.

Rosenblueth, A., Wiener, N., and Biglow, J. H. (1943). Behavior, purpose, and teleology. *Phil. Sci.* **10**, 18–24.

Simpson, G. G. (1963). Biology and the nature of science. *Science,* **139**, 81–88.

Simpson, G. G. (1967). The crisis in biology. *Amer. Scholar,* **36**, 363–377.

Wiener, N. (1948). "Cybernetics." MIT Press, Cambridge, Massachusetts. Rev. ed. Wiley, New York, 1961.

Yamamoto, W. S., and Brobeck, J. R. (eds.) (1965). "Physiological Controls and Regulations." Saunders, Philadelphia, Pennsylvania.

Flow Processes in the Steady State || *Chapter 2*

Introduction   2.1

An organism is continually exchanging material with its environment, taking in food and oxygen, and discharging the waste products. The materials consumed are transported to various sites in the body where they are transformed into other molecules and then stored for future use, or excreted. Ingestion, transport, exchange, and excretion will be termed *flow processes,* and usually may be represented by the flow of material or fluid into and out of a *tank, compartment,* or *capacity.* For the present, these three terms will be used interchangeably.

This chapter is concerned with the flow or transport of material through a series of processes when that flow has reached a *steady state,* here defined as the condition in which all flow rates and related quantities have reached constant values. From a physical standpoint, an organism is in the steady state if its individual processes and the entire system maintain a behavior that does not change with time. A prerequisite for the steady state is usually a constant environment, together with no changes in the internal processes.

It may be objected that the steady state is a somewhat artificial condition, and does not characterize the organism as it normally responds to a changing environment. The point is well taken, although the steady state is not unphysiological and may be attained under proper experimental conditions. Furthermore, even though the steady state may not be achieved in practice, it is essential that the implications of this concept be clear before attempting a discussion of dynamic behavior.

The concept of a steady state should be clearly distinguished from that of *equilibrium.* In the thermodynamic sense, equilibrium exists only when there is no flow of energy or material; that is, it is equivalent to the terminal state, death. On the other hand, life is characterized by a continuing flow of material and energy, and a steady state is reached if all possible disturbing factors remain constant. The fact that in life new disturbances appear before the effects of previous ones have had a chance to disappear does not invalidate the concept. See Section 6.3 for an account of the sinusoidal steady state.

The choice of fluid flow for expository purposes is motivated by two considerations. First, the mathematical relations describing fluid flow as we will develop them are identical with those for many other physical processes. Second, fluid flow is a readily visualized process.

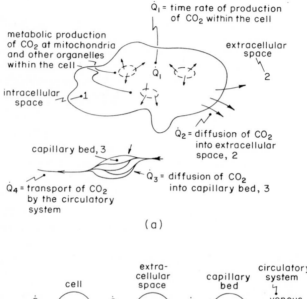

$\dot{Q}_1$ = time rate of production of CO_2 within the cell

metabolic production of CO_2 at mitochondria and other organelles within the cell

extracellular space

intracellular space

$\dot{Q}_1$

1

2

$\dot{Q}_2$ = diffusion of CO_2 into extracellular space, 2

capillary bed, 3

$\dot{Q}_3$ = diffusion of CO_2 into capillary bed, 3

$\dot{Q}_4$ = transport of CO_2 by the circulatory system

(a)

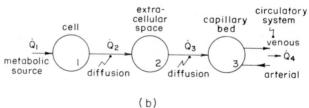

extra-cellular space

capillary bed

circulatory system

cell

$\dot{Q}_1$

$\dot{Q}_2$

$\dot{Q}_3$

venous

metabolic source

1

2

3

$\dot{Q}_4$

diffusion

diffusion

arterial

(b)

Fig. 2-1. Flow of CO_2 by diffusion from a metabolizing cell to the venous circulation. Q is a suitable measure of quantity, and $\dot{Q}$ is the time rate of flow, or quantity per unit time (see Section 2.4). (a) CO_2 flow from cell through extracellular space into the venous circulation. (b) Flow graph with three compartments. In the steady state, $\dot{Q}_1 = \dot{Q}_2 = \dot{Q}_3 = \dot{Q}_4$.

a. A large portion of the following discussion concerns the flow of materials (or energy) into and out of specific regions or compartments in the organism. The concept of a compartment (to be defined subsequently) permits one to divide a complex system into a number of subsidiary processes for ease in analysis. This figure shows the production of CO_2 in a metabolizing cell and the flow of CO_2 from the cell into the extracellular space, from there into the capillaries, and by transport in the circulatory system, to the lungs. Each of these regions may be considered a compartment.

b. The cellular membrane is permeable to CO_2 so that this molecule may diffuse in either direction, the net flow depending upon the concentration difference in the two compartments, or the concentration gradient. The time rate of flow, $\dot{Q}_2 = dQ_2/dt$, is directly proportional to the difference in CO_2 concentration in compartments 1 and 2, and implies a higher concentration in 1 than 2. Further definitions will be found in Section 2.4.

c. A steady state will be attained when, for a constant production rate $\dot{Q}_1$, all other flow rates reach the same magnitude. Thus, considering the series of compartments from cells to lungs, each compartment in sequence must have a somewhat lower CO_2 concentration than its predecessor in order for there to be a constant flow of CO_2 through all compartments at the same rate as it is produced within the cell. Inasmuch as the resistance to diffusion between the several compartments varies through the system, the difference in concentration between each pair of adjacent compartments will be greatest where the diffusion resistance is the greatest.

d. The diffusion of CO_2 into the capillaries differs somewhat from the diffusion out of the cell. The arterial supply to the capillary bed is deficient in CO_2, but the blood acquires CO_2 on its passage through the capillaries, so that the CO_2 concentration in the venous blood is essentially in equilibrium with that in the extracellular space. Once the CO_2 reaches the venous circulation ($\dot{Q}_4$), its movement is set by the blood velocity and not by a concentration gradient. The latter again becomes a factor when the blood reaches the alveoli and CO_2 diffuses into the lungs.

e. Molecules produced by chemical action within a compartment may be treated as an input flow, as was done with CO_2 in this example. However, the input to the first compartment may come from some other system.

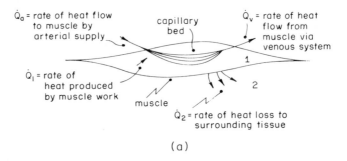

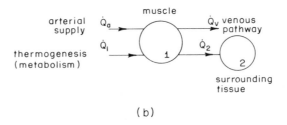

Fig. 2-2. Muscle considered as a homogeneous thermal compartment. The heat flow rates, $\dot{Q}$, are measured in calories per unit time. (a) Heat flow pathways showing heat sources $\dot{Q}_a$ and $\dot{Q}_1$, and the heat losses $\dot{Q}_v$ and $\dot{Q}_2$. (b) Flow graph. In the steady state, $\dot{Q}_a + \dot{Q}_1 = \dot{Q}_v + \dot{Q}_2$. Note that these flow rates may have quite different values, depending on the state of muscle contraction.

a. The flow of energy (heat, for example) follows mathematical relations that are quite similar to those for material flow, when suitable simplifying assumptions are permitted. As an example, consider the flow of heat in an active muscle.

b. The metabolic heat produced by a muscle (thermogenesis) will be a minimum in the relaxed or resting state, and a maximum during contraction. As a consequence of the heat produced, the muscle temperature must rise so as to increase the heat losses to a point where the production and loss of heat have equal rates. When this occurs, the muscle temperature remains constant, and a thermal steady state exists.

c. There are two possible heat sources (inputs) for the muscle: heat produced by metabolism within the muscle, and heat transported to the muscle by the arterial system.

d. Of the two possible avenues of heat loss, the first is via the circulatory system, since if the muscle heats the circulating blood (the converse of the process described in the previous paragraph), the venous blood will be at a higher temperature than the arterial supply. A second avenue of heat loss is by tissue conduction, which carries heat to adjacent tissues and from there to the overlying skin and surrounding air.

e. Although a muscle and its capillary network are by no means a homogeneous and isotropic structure, it is possible, for certain purposes, to consider them as such, and speak of a thermal compartment. An idealized thermal compartment would be characterized by a uniform internal temperature and certain heat inputs (sources) and outflows (losses or sinks).

f. A *compartment* may be defined in a number of ways:

(1) A physically defined region or location having a boundary across which material (or energy) moves at a measurable rate, so that the material inside and that outside are readily distinguishable (Riggs, 1963).

(2) A compartment can also consist of a given kind of molecule or chemical form, regardless of where it may be located in the body (such as hemoglobin).

(3) An anatomical, physiological, chemical, or physical subdivision of a system throughout which the concentration of a given substance is uniform at any given time (Brownell, Ashare, 1967).

g. The concept of compartment, although extremely useful in biology, is an abstraction whose validity must be assessed in each instance.

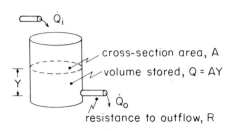

Symbol	Definition	Typical units (cgs)	Symbol	Definition	Typical units (cgs)
Q	Quantity of material Volume of material	cm^3	A	Cross-section area of tank	cm^2
$\dot{Q}$	Time rate of flow Time rate of change	cm^3/sec	R	Hydraulic flow resistance	$gm\ sec/cm^5$
Y	Height of liquid above outlet	cm	δ	Density	gm/cm^3

Fig. 2-3. Symbols and definitions for fluid flow through a compartment. The storage function is obtained from the conservation of mass in the following manner:

$$\text{inflow rate} - \text{outflow rate} = \text{time rate of accumulation}$$
$$\dot{Q}_i \quad - \quad \dot{Q}_0 \quad = \quad A(dY/dt)$$

a. The tank (compartment) in this figure illustrates certain basic relations common to compartmental models of biological processes. The material influx occurs at the rate $\dot{Q}_i$, and the efflux at the rate $\dot{Q}_0$. The volume stored is $Q = AY$.

b. The quantity of material can be measured by its volume, weight, or mass, as may be convenient. It can also be expressed in moles, or in some measure of concentration. In the examples to follow, quantity will be expressed in volume units, but the concepts and relations discussed are readily transferable to systems using other measures of quantity (Riggs, 1963, Chapter 2).

c. Although cgs or mks units are suggested for the several quantities defined in the figure, a single set of units should be selected for a given process. Each equation describing a physical process must be dimensionally homogeneous, that is, all terms in a given equation must have the same dimensions. However, the several equations describing a given process need not all have the same dimensions and in fact will usually have different dimensions. Moles and cubic centimeters cannot be added, but a certain system might well use both units, each to describe a different relationship within the system.

d. The dot notation ($\dot{Q}_0$) is equivalent to the first derivative with respect to time, dQ_0/dt, where Q_0 is the volume or mass leaving the tank, and $\dot{Q}_0$ is thus the time rate of flow in suitable units of quantity per unit time. If $\dot{Q}_0 = \dot{Q}_i$, the process is in a steady state for which both Q and Y are constant.

e. A compartmental process has two aspects: the storage or accumulation of material within the compartment, and the loss from it. A quantitative description of this flow process requires two mathematical functions, the storage and loss functions, which must be derived from the applicable physical laws or obtained by experiment. Note that the term function is used here in a mathematical, and not a physiological, sense.

f. The *storage function* relates the quantity of material stored within the compartment, or rate of change of this quantity, to the inflow and outflow rates. This relation is usually derived from a conservation law – in the case shown, from the conservation of mass. The quantity of material may also be expressed in suitable units of concentration. Note that the equation in this figure is quite general in that it is valid when the flow rates $\dot{Q}_i$ and $\dot{Q}_0$ are changing or when they are constant. In the steady state, $\dot{Q}_i = \dot{Q}_0$, and $A(dY/dt) = 0$.

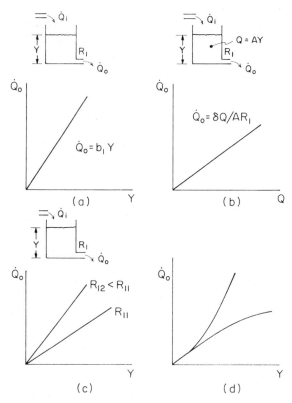

Fig. 2-4. Possible loss functions associated with fluid flow in a compartment. (a) Loss as a function of hydrostatic head. (b) Loss as a function of quantity stored. (c) Loss for different values of outflow resistance. (d) Nonlinear loss functions.

2 FLOW PROCESSES IN THE STEADY STATE

a. The manner in which material is removed or leaves a compartment is subject to a variety of physical laws, and the loss function may therefore take a number of forms. A few of the possibilities are considered here.

b. The variety of possible loss functions may be seen from the following partial list of loss mechanisms: (1) Loss by fluid flow is dependent upon the fluid flow resistance. (2) Loss by diffusion is set by the membrane permeability. (3) Thermal loss by heat flow depends upon conduction, convection, and radiation. (4) Loss as a consequence of chemical reaction depends upon the applicable reaction rate mechanisms.

c. The present discussion is restricted to the case of fluid flow through a constant outflow resistance. The loss function may be found in three different ways: (1) Using a phenomenological or experimental approach, one might observe that $\dot{Q}_0$ was linearly proportional to Y as in part (a) of the figure. That is, $\dot{Q}_0 = b_1 Y$ where b_1 is a constant. (2) A second approach makes use of basic physical relations. It is assumed in this example that $\dot{Q}_0$ depends upon the hydrostatic head, δY, and the flow resistance, R_1, as given by $\dot{Q}_0 = \delta Y / R_1$. The hydrostatic head δY is the same as the hydraulic pressure available at the outlet to drive fluid through R_1. It may also be considered as the potential "force" that produces the outflow. (3) A third viewpoint may be obtained by noting that since $Q = AY$, one can write $\dot{Q}_0 = \delta Y A / A R_1 = \delta Q / A R_1$. Thus, an outflow of this nature might be looked upon as a loss proportional to the quantity stored. This is suggested by part (b) of the figure.

d. A decrease in R_1 will obviously increase the outflow $\dot{Q}_0$, as in (c). The double subscript notation used here employs the first subscript to designate a specific variable, and the second subscript to designate a specific value for that variable. Double subscripts will be employed for a different purpose in later discussions.

e. A loss function for which the outflow is proportional to the quantity stored is similar to the *law of mass action* in chemical kinetics. Thus, the relations developed for these fluid flow models will be applicable to processes for which the mass action law is valid, or any other process having similar functional relationships.

f. Although not shown in this figure, the loss could occur at a constant rate independent of the input and the quantity stored, or could be a function of the lifetime of the particles within the compartment. A variety of nonlinear loss functions are also conceivable, as in (d).

2.5   LOSS FUNCTIONS

15

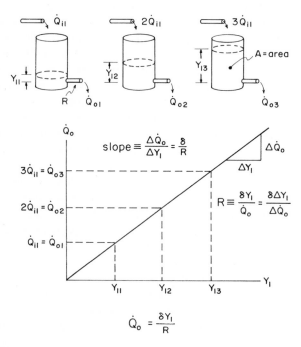

Fig. 2-5. Single compartment flow process in three different steady states produced by changes in the inflow rate. Note that the quantity of material stored is proportional to the inflow rate.

　　　　　　　　2　FLOW PROCESSES IN THE STEADY STATE

a. The three different steady states in this figure are the result of different values of the inflow $\dot{Q}_{i1}$. The second subscript denotes a specific value of $\dot{Q}_i$, and for convenience, the flows shown are multiples of $\dot{Q}_{i1}$. The steady state is reached when $\dot{Q}_0 = \dot{Q}_i$, and this results in three values of Y_1.

b. The functional relation between $\dot{Q}_0$ and Y_1 is depicted by the straight line on which the three steady-state operating points are indicated. The outflow resistance is the ratio $\delta Y_1/\dot{Q}_0$, which remains constant for all flow rates. The ratio can also be expressed in terms of incremental quantities ΔY_1 and $\Delta\dot{Q}_0$. Another fluid having a different density δ would therefore be represented by a line of different slope.

c. The *slope* of the line, defined as shown, is inversely proportional to R, and the constant of proportionality is δ. Since the slope of this line is everywhere the same, the parameter R is a constant. This shows that $\dot{Q}_0$ is proportional to Y_1 over the range of Y_1 shown. This model is said to be *linear*. Not all steady-state characteristic curves exhibit such a linear relationship (see Sections 2.5 and 2.12).

d. The model shown in this figure could represent a variety of physical processes in which a flow or transformation takes place at a rate proportional to an intensity factor, or a quantity stored. An example is the diffusion of particles through a thin membrane when the rate of movement is proportional to the particle concentrations. Another example is a chemical reaction with a rate proportional to the number of reacting particles (the law of mass action).

e. The values attained by Y_1 and $\dot{Q}_0$ in the steady state will depend upon both $\dot{Q}_i$ and R, and any change in these quantities will affect the steady-state operating point. The cross-section area A does not affect the steady-state values of Y_1 or $\dot{Q}_0$, but does change the quantity stored, $Q = AY_1$.

f. It is quite conceivable that with a large $\dot{Q}_i$, the tank would overflow. Practically all physical systems exhibit limitations that do not appear in their mathematical description. The functional relations described here and in the following sections hold only within the physical limitations of the process.

g. This figure says nothing regarding the manner in which the steady state is attained. Clearly, if the tank were empty when $\dot{Q}_{i1}$ was established, Y_1 would rise to the value Y_{11}. Similar statements may be made for any other change, the value of Y_1 increasing or decreasing until $\dot{Q}_0 = \dot{Q}_i$. The dynamic behavior is discussed in Chapter 4.

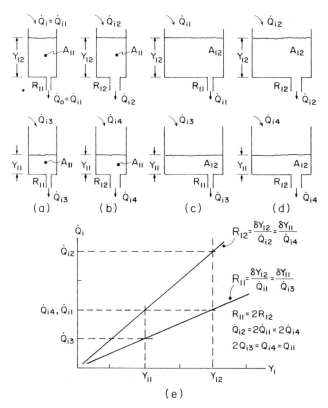

Fig. 2-6. Four different flow processes consisting of combinations of A_{11} or A_{12} and R_{11} or R_{12} are shown in (a), (b), (c), and (d). Each process appears in two steady states produced by different values of $\dot{Q}_i$. (e) Steady-state characteristic curves for the several processes with the operating points indicated.

2   FLOW PROCESSES IN THE STEADY STATE

a. Once the assumption is made that a process can reach a steady state without exceeding physical limitations, it then becomes of interest to ascertain the manner and extent to which the physical parameters determine the steady-state relations. The figure shows four different processes, (a), (b), (c), (d), and each is shown in two distinct steady states.

b. The factors which may contribute to the steady state are the inflow $\dot{Q}_i$ and the quantities R, A, and δ. It is assumed for this figure that δ is the same for all cases shown. Two arbitrary values have been assumed for A and R, yielding four distinct processes, and the two different values assumed by $\dot{Q}_i$ for each process yield the eight cases shown, all of which are in a steady state. Note that although cases (a) and (b) appear to be in similar steady states, the comparison is without much significance because the processes are not the same ($R_{11} > R_{12}$).

c. In a given physical system there is usually some choice in the variables used to describe its behavior. The previous discussion employed Y and $\dot{Q}_0$ as the dependent variables, although the quantity Q could just as well have been used instead of Y. Since $Q = AY$, the variables Y and Q are linearly related, and the relations obtained using either choice are the same except for a scale change. The choice of variable in a specific problem will be governed by the nature of the question under study, and by the matter of convenience.

d. The tank area A, while not a factor in determining the steady-state magnitude of Y, is linearly related to Q, and therefore does change the steady state.

e. It is conceivable that the seven quantities defined for this process ($\dot{Q}_i, \dot{Q}_0, Q, A, Y, R, \delta$) could all be changing at a given instant, and be interrelated. However, it is more likely that A, R, and δ are constant in a given physical situation, although they may well take on quite different values from one problem to the next. Those quantities that remain constant in a given problem are termed *parameters*. The remaining quantities, Q, Q_i, Q_0, and Y, which may change during some period of time under study, are termed *variables*, among which we note two classes: the *independent* and *dependent* variables. The independent variables ($\dot{Q}_i$) are those quantities whose magnitudes are independent of what goes on in the process; they are frequently termed *inputs*. The dependent variables (Q, Y, $\dot{Q}_0$) are functions of the inputs, the parameters, and of each other. They may be termed *outputs*.

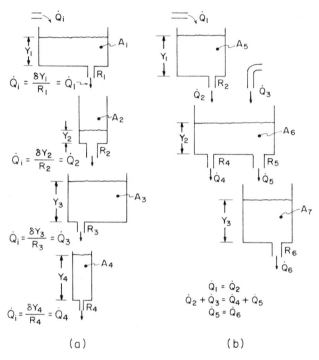

$$\dot{Q}_i = \frac{\delta Y_1}{R_1} = \dot{Q}_1$$

$$\dot{Q}_i = \frac{\delta Y_2}{R_2} = \dot{Q}_2$$

$$\dot{Q}_i = \frac{\delta Y_3}{R_3} = \dot{Q}_3$$

$$\dot{Q}_i = \frac{\delta Y_4}{R_4} = \dot{Q}_4$$

(a)

$$\dot{Q}_1 = \dot{Q}_2$$
$$\dot{Q}_2 + \dot{Q}_3 = \dot{Q}_4 + \dot{Q}_5$$
$$\dot{Q}_5 = \dot{Q}_6$$

(b)

Fig. 2-7. Multicompartment systems, in which all processes are noninteracting, are shown in the steady state. (a) The four compartments are arranged in series, with no intermediate inflows or outflows. (b) Three noninteracting compartments arranged in series, but with an additional inflow and outflow to the second compartment.

2 FLOW PROCESSES IN THE STEADY STATE

a. If a series of single-capacity flow processes are arranged as in (a), so that the outflow of one becomes the inflow to the next, then each compartment may be treated in the manner previously described, and in the steady state, $\dot{Q}_i = \dot{Q}_1 = \dot{Q}_2 = \dot{Q}_3 = \dot{Q}_4$. Conservation of mass applies to the entire system as well as to the individual portions of it.

b. To preserve the same flow through each of the compartments, the liquid levels will assume values proportional to the individual resistances. Thus, a change in any resistance will change the magnitude of Y in that tank, but will not affect the steady-state flow through any tank, or through the entire system.

c. Compartment size, as represented by the area A, has no relation to the magnitude of Y, although it does affect the quantity stored, Q. In this system a new steady state as regards $\dot{Q}_4$ can only occur if there is a change in $\dot{Q}_i$.

d. The system of four compartments described above is termed *noninteracting* to denote the fact that each tank may be treated individually. The flow from one compartment to the next is not dependent upon conditions in the following compartment.

e. The concept of noninteraction is a means of describing the *coupling* between two compartments. The term *coupling*, as used here, refers to the manner in which one process affects the flow between adjacent compartments. One could say that the first tank is coupled to the second by virtue of the flow $\dot{Q}_1$, whose magnitude depends on Y_1, but not at all on conditions in the second tank. This type of coupling is termed *unilateral* or *noninteracting*; that is, there is a flow from the first to the second tank, but not in the opposite direction, and tank two has no influence upon material flowing into itself.

f. For each value of the input $\dot{Q}_i$, the system will reach a unique steady state, assuming only that the physical limitation of tank capacity is not exceeded. The term *unique steady state* simply means that all dependent variables take on unique values for each value of the input. Not all systems exhibit this property.

g. The three compartments shown in (b) are again coupled in a non-interacting manner, but in this instance there are two inflows ($\dot{Q}_1$, $\dot{Q}_3$) and two outflows ($\dot{Q}_4$, $\dot{Q}_6$) to the system. The principle of conservation of mass is now applied to each compartment as well as to the entire system.

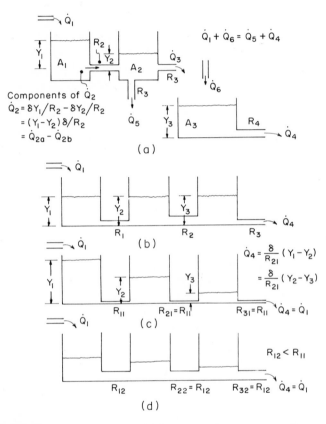

Fig. 2-8. (a) Three-compartment system in which the first two compartments are coupled in an interacting manner so that the flow $\dot{Q}_2$ is proportional to the difference $Y_1 - Y_2$. (b) Interacting processes in equilibrium, but not a steady state. (c) Steady state. (d) Steady state, but with smaller values for the flow resistances.

2   FLOW PROCESSES IN THE STEADY STATE

a. The first two tanks in (a) are coupled in an *interacting,* or *bilateral,* manner, whereas the coupling between the second and third tanks is non-interacting, or unilateral. These statements refer to the fact that $\dot{Q}_2$ depends upon the conditions in both tanks whereas $\dot{Q}_3$ depends only on the second.

b. The bilateral coupling between the first two compartments may be viewed in two different but equivalent manners. The net flow to the right ($\dot{Q}_2$) depends linearly upon the difference $Y_1 - Y_2$. The flow $\dot{Q}_2$ can thus conceivably be in either direction or zero. This is a physical statement describing the coupling, and is not inconsistant with the fact that in this system Y_2 will not exceed Y_1.

c. The flow $\dot{Q}_2$ may also be viewed as the resultant of two components, one a flow $\dot{Q}_{2a}$ from tank 1 to 2, and the second component $\dot{Q}_{2b}$ from tank 2 to 1.

d. In the steady state, the conservation of mass and the loss function jointly determine the value of each Y. Thus, (b) cannot be a steady state because $Y_1 = Y_2 = Y_3$, and although the state is physically possible, it cannot persist. On the other hand, in the steady state, $Y_1 > Y_2 > Y_3$, and the differences between the levels will depend upon the flow rate and the several R's.

e. The bilateral coupling described above is symmetrical in that the magnitude of R is independent of the direction of fluid flow. However, there are many instances in which this symmetry of coupling is not observed and the magnitude of R (or the equivalent coupling term) is quite different in the two directions. This would produce a bilateral but *asymmetrical* coupling, an extreme example being that of unilateral coupling with no flow possible in the reverse direction.

f. No fundamental difference can be found in comparing interacting and noninteracting processes, since both exhibit the same kinds of behavior. However, when it comes to the analysis of system response, there is a considerable difference in the ease with which calculations can be made, in favor of the noninteracting processes. Fortunately a large number of the processes appearing in homeostatic systems are of the noninteracting variety.

g. Examination of the coupling between the component processes in a physical system (mechanical, electrical, chemical, neural, and hormonal) will show that the several types of coupling described above are found quite generally. However, bilateral coupling should not be confused with feedback, which is defined explicitly on p. 89.

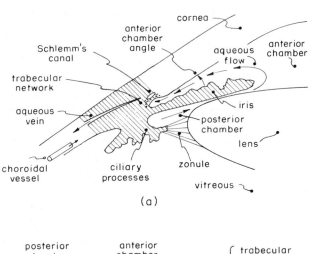

(a)

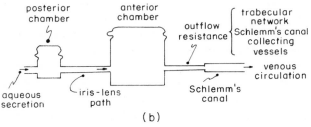

(b)

Fig. 2-9. Aqueous flow through the anterior segment of the eye. (a) Anatomical diagram showing the path of aqueous from the ciliary process, through the posterior and anterior chambers, and into Schlemm's canal and the aqueous vein. (b) Two-compartment model of this flow process (Becker and Shaffer, 1961).

a. Aqueous humor fills the posterior and anterior chambers of the eye, and since aqueous is being continuously secreted by the ciliary processes, there is a continuous flow through the eye and into the aqueous veins. The anatomical diagram in (a) gives some idea of the geometry of this pathway.

b. This flow system differs from the models previously described in that the compartments representing the posterior and anterior chambers are closed structures, but have some elastic properties as suggested by the schematic diagram in (b). Thus, as the aqueous humor accumulates within the two chambers, the pressure rises proportionately. The flux of aqueous between compartments is assumed to be proportional to the pressure difference. This linear assumption appears to be adequate for most purposes.

c. The flow will reach a steady state if the secretion rate is constant; then the pressure differentials between compartments are just sufficient to maintain this flow rate.

d. The secretion of aqueous is an active process, the rate of which depends upon neural and hormonal factors as well as arterial pressure. Either hypersecretion or increased resistance in the pathway will lead to excessive pressures throughout the system, with resultant pain and deterioration (glaucoma).

e. Changes in the flow resistance may occur at a number of points. A reduction in the cross-section area for flow between the iris and lens leads to *pupillary block*. The angle of the anterior chamber may narrow (angle closure glaucoma). Although the trabecular network provides a flow resistance that is presumably under neural control, congenital and pathological factors may increase the flow resistance leading to *primary open-angle glaucoma.*

f. Pressures within the vascular system may affect both the secretory process and the final disposition of the aqueous by flow into the collecting veins.

g. The use of only two compartments for this flow model is largely based upon the identification of two obvious compartments: the posterior and anterior chambers. However, careful experimental studies reveal that aqueous also moves into the iris and lens so that two additional compartments may be required for some models.

h. There is considerable evidence to support the hypothesis that the intraocular pressure is regulated by negative feedback pathways, although none are shown in this figure.

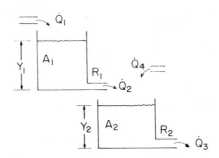

Disturbances						Consequences			
R_1	R_2	A_1	A_2	$\dot{Q}_1$	$\dot{Q}_4$	$\dot{Q}_2$	$\dot{Q}_3$	Y_1	Y_2
0	$+\Delta$	0	0	0	0	0	0	0	$+\Delta$
0	0	$+\Delta$	0	0	0	0	0	0	0
$+\Delta$	0	0	0	0	$-\Delta$	0	$-\Delta$	$+\Delta$	$-\Delta$
0	0	0	0	$+\Delta$	0	$+\Delta$	$+\Delta$	$+\Delta$	$+\Delta$

Fig. 2-10. The consequences for the steady state of a fluid flow system are shown for a variety of imposed disturbances. The table shows single and multiple disturbances, along with the resulting changes in the dependent variables $\dot{Q}_2$, $\dot{Q}_3$, Y_1, Y_2, where 0 indicates no change, $+\Delta$ is a positive increment, and $-\Delta$ is a negative increment.

2   FLOW PROCESSES IN THE STEADY STATE

a. In a biological system, one finds almost continual changes in the environment, and in the system parameters themselves. These disturbances will generally (but not always) affect the steady state, and cause the system to move to a new operating point. For example, in these fluid flow systems, the governing relation is the conservation of mass, so that a disturbance will change the steady state if such is required to preserve the conservation relations.

b. The disturbances in the first three rows of the table affect the system parameters and thus result in a "new" system. On the other hand, the disturbance in the fourth row affects only the input, and the system parameters remain unchanged.

c. The disturbance in the second row results in no change in the four variables shown, but there will be an increase in another dependent variable, $Q_1 = A_1 Y_1$.

d. With each disturbance, there will be associated transient changes as described in the next chapter. Note, however, that the behavior of the system following a disturbance will be that of the disturbed system, and could possibly be quite different from that of the original (undisturbed) system.

e. As a general rule, it is to be expected that physical disturbances will appear that affect each and every term in the defining equations of the system. Moreover, disturbances can change the form of the system equations themselves and thus affect all aspects of system response and behavior.

f. As an example, consider disturbances to the respiratory system. An increase in muscular work will increase the production rate of CO_2 and produce an increase in the ventilatory rate. Fibrosis of the lungs results in a reduction in lung volume. Emphysema or asthma increases the resistance of the air passages.

g. The fluid flow processes discussed up to this point may be considered as *analogs* of a variety of physical processes having similar storage and loss functions. Two physical systems are said to be analogous if they behave in identical manners when disturbed in the same way. This means that the defining equations are of identical form, and two systems can be shown to be analogous only by establishing this fact. The above definition of analog is somewhat restrictive, compared to those found in much of the current literature. However, the analogous relation thus becomes precise, and so it is in this sense that the term will be used here. It should also be noted that an analog is not unique, in the sense that a given fluid flow analog may represent many different physical systems.

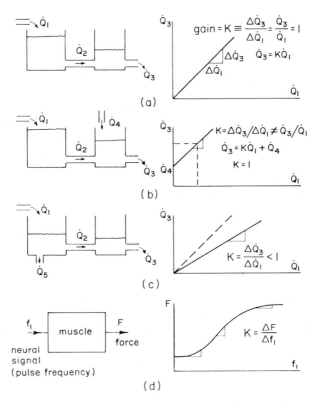

Fig. 2-11. Steady-state transmission (gain) for systems with (a) single input and single output, (b) two inputs and one output, (c) single input and two outputs. (d) Signal transmission through a muscle considered as a transducer.

a. In the systems previously described, input and output flow rates have been identified, and if a conservation law was applicable, these two rates in the steady state were either equal or bore a fixed relation to each other. In other than flow processes, the input and output variables may be quite unlike quantities, and the application of a conservation law may not be apparent.

b. In (a), the conservation of mass leads directly to the equality of $\dot{Q}_1$ and $\dot{Q}_3$ in the steady state. This will be true for all values of $\dot{Q}_1$, providing only that the physical limits of the compartments are not exceeded.

c. The relation between the input and output quantities is described mathematically by the *transmission* or *transfer function*. The present discussion is limited to the steady state, and the transfer function is then termed the *gain* of the process. Gain (represented by K) is defined as the ratio of steady-state increments in output and input variables, $\Delta\dot{Q}_3/\Delta\dot{Q}_1$, where the increments are normally small changes. For clarity, this ratio may be termed the *incremental* or *small signal gain*. The term "gain" is borrowed from engineering and amplifier technology, and while possibly not too appropriate for a biological process, its welcome brevity commends its continued usage. The "gain" may be greater or less than unity.

d. With additional inputs and outputs [$\dot{Q}_4$ in (b), and $\dot{Q}_5$ in (c)] the steady-state characteristic, although still a straight line, is displaced as shown.

e. For those characteristics that are straight lines, the output is said to be a *linear function* of the input. In (a), the gain K is everywhere unity, but in (b), although K is unity, there is a component in the output due to $\dot{Q}_4$ and not related to $\dot{Q}_1$. In (c), the gain is less than unity, but in other examples it may be greater.

f. The muscle, shown simply as a block in (d), differs from previous examples in having an input and output that are different kinds of physical quantities. Such a component, termed a *transducer,* is defined as one whose input and output variables are unlike physically. A transducer couples two different physical systems, in this case the nervous system and mechanical components of the skeleton. In general, biological receptors and effectors are transducers having a gain expressed in suitable dimensions.

g. In (d), the incremental gain (force/pulses per second) varies with the magnitude of f_1, and the transducer is said to be *nonlinear*. However, over the intermediate range (and frequently the physiological range) the gain may be sufficiently constant to permit treating the component as a linear device for small signals.

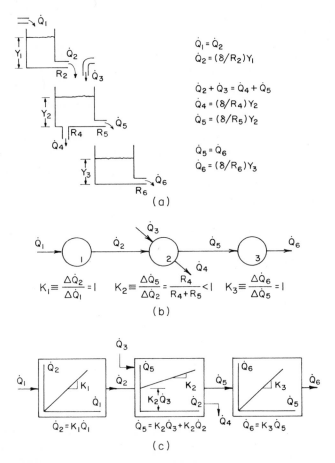

Fig. 2-12. Analysis of the steady state of a three-compartment system. (a) The three processes and their related steady-state equations. (b) Flow graph, and (c) block diagram with relations derived from the process equations in (a).

2 FLOW PROCESSES IN THE STEADY STATE

a. The system of flow processes at the top of this figure is the same as that shown previously in Fig. 2-7b. Note that whereas the first and third compartments have only a single input and single output, the second compartment has two of each.

b. The *flow graph* in (b) represents the same three compartments together with all the flows. The circle is chosen to represent a compartment to denote the fact that it satisfies a conservation law, from which the relations between the several flows have been obtained. Although the heavy lines signify material flow, the flows may also be termed signals, as described further in Section 2.17. Only the steady-state relations are shown in this figure.

c. The *block diagram* in (c) provides another way of representing this system in which not only flows but any other physical quantity may serve as the input and output variables. The relations between the variables may be given in either graphical or algebraic form (Riggs, 1970).

d. From the standpoint of the whole organism, only $\dot{Q}_1$ and $\dot{Q}_6$ may be of interest, and the additional flows in the second compartment may appear as constants or parameters in the overall relation. Thus, a term dependent upon $\dot{Q}_3$ appears in the expression for $\dot{Q}_5$, but $\dot{Q}_4$ is buried in the constant K_2.

e. The first and third compartments, each having only a single input and output, have characteristics with unity slope, that is, $K = 1$. This follows directly from the conservation law.

f. The block diagram treats all variables as signals and a conservation law, if applicable, is buried within the defining relations. As a consequence, the output signal from a block may be directed to any number of other blocks without regard for conservation of signal. On the other hand, the flow diagram represents compartments and must be drawn so that the input and output signals are conserved.

g. The flow graph is reasonably isomorphic to the physical system it describes in that most of the physical variables can be identified on the graph, whereas this is usually not the case with a block diagram. The latter should be viewed as a means of representing the functional relations between the several variables in a mathematical or graphical form, and since these mathematical relations may be written in many different ways, the block diagram may bear little resemblance to the physical system even though it does depict the correct mathematical relations.

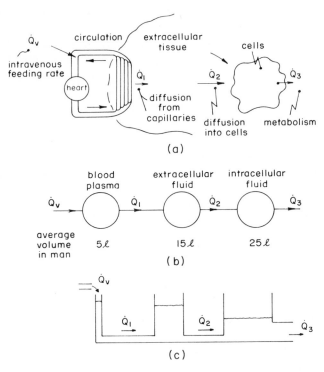

Fig. 2-13. Intravenous glucose feeding as an example of a three-compartment system. (a) Schematic diagram of the flow processes. (b) Flow graph. (c) Fluid flow model showing the approximate relative sizes of the three compartments.

2   FLOW PROCESSES IN THE STEADY STATE

a. Intravenous feeding results in glucose transport by the circulatory system into the extracellular space and from there into the cells. The transport from plasma to extracellular fluid and then into the individual cells occurs by diffusion. Under normal conditions, although the kidney filters out the glucose from the blood, it is almost entirely reabsorbed so that there is no net loss of glucose by renal function, and the kidneys are therefore not shown.

b. The figure assumes that the intravenous feeding rate is below the maximum metabolizing rate of the cells, so that all the glucose is metabolized. Furthermore, a number of processes that normally serve to regulate the blood glucose level have been omitted.

c. The hydraulic analog (c) has been proportioned to show the relative sizes of the three compartments in man (b). In addition, the three flow "resistances" must correspond to the diffusion "resistances" in the prototype.

d. Inasmuch as these individual compartments are coupled in an interacting manner, the net flow at any point depends upon the difference in concentrations of glucose in the two adjacent compartments. The steady state is reached when $\dot{Q}_v = \dot{Q}_1 = \dot{Q}_2 = \dot{Q}_3$. An increase in $\dot{Q}_v$ above the maximum value of $\dot{Q}_3$ may well lead to hyperglycemia.

e. The movement of glucose occurs by two different kinds of processes; namely, *passive movement* by diffusion across the capillary and cell membranes, and *active transport* by the circulatory system. Active transport in this instance is provided by the blood flow, which carries glucose throughout the body tissues at a rate dependent upon the heart activity. This circulatory transport is not shown explicitly in the diagram, and although it adds a transport time to the overall process, this affects the dynamic, but not the steady state, behavior.

f. In those examples in which passive diffusion occurs, the interacting tanks shown in this model serve as an adequate analog. However, for some forms of active transport and for a variety of other biological processes, the noninteracting coupling used in previous examples is required.

g. Although in the process illustrated in the figure, the flow is normally toward the right, under transient conditions and especially in the presence of feedback (not shown) the individual flows may reverse momentarily. The model is still valid, providing the physical counterpart of the resistance is a constant, and independent of the *direction* of flow.

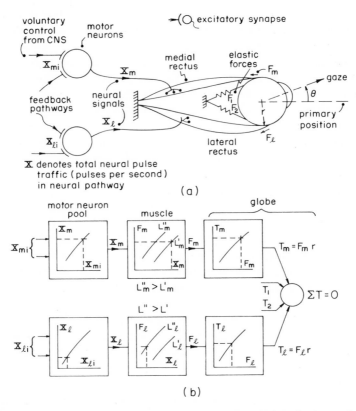

Fig. 2-14. Control of the position of one eye in the horizontal plane by the two horizontal recti muscles. (a) Anatomical diagram in which the feedback pathway for visual signals has been omitted. (b) Block diagram showing the two pathways providing torques of opposite sign to the eyeball.

2   FLOW PROCESSES IN THE STEADY STATE

a. Horizontal movement of the eye is brought about by the two horizontal rectus muscles, which provide torques of opposite sign to the globe. A steady state occurs when the eyes fixate a target and the position of each eye remains constant.

b. To maintain a fixed gaze at, say, an angle θ from the primary position, the sum of all torques acting on the globe must equal zero, and this in turn requires a precise balance in the muscle actions. To achieve this balance, feedback signals from the visual fibers and muscle proprioceptors are required. This feedback is not shown explicitly, but only as providing an additional input to the motor neurons.

c. The group of motor neurons which send signals to one muscle, termed the *motor neuron pool,* is represented by one block. Each pool has thousands of input and output signals, although these are shown in this figure as single pathways. At any time only a portion of the nerve cells in the pool may be active, and these may be firing at a variety of rates.

d. The variable X is a measure of the total pulse traffic on all fibers in a given neural pathway. The X_{mi} and X_{li} represent the neural signals flowing into the motor neuron pool; the X_m and X_l are the signals in the motor nerves to the muscles. These latter signals from the pools are a consequence of voluntary signals from the CNS, as well as feedback signals, and the total input to the pools must be such as will maintain the torque balance required for steady gaze.

e. The force produced by a muscle depends on the intensity of the neural signals (X) and also upon muscle length (L). The force increases with length up to some maximum value, and this relationship is shown by the two curves drawn in each of the muscle blocks for different values of L.

f. The right hand blocks serve to show the conversion of muscle force into torque acting on the globe. The effective radius r may change with θ.

g. Passive elastic torques are associated with the orbital tissues. Although two such torques are shown, one acting in each direction, under steady-state conditions there will be only one elastic restoring torque.

h. In this neuromuscular system, individual components and biological processes are considered as performing transformations from one variable to another. The character of these relations must be obtained from physical reasoning, or more likely, from experiment.

2.15 CONTROL OF THE EXTRAOCULAR MUSCLES

Conservation law | Loss functions

$$\dot{Q}_1 = \dot{Q}_2 \qquad \dot{Q}_2 = (\delta/R_2)Y_1$$
$$\dot{Q}_2 + \dot{Q}_3 = \dot{Q}_4 + \dot{Q}_5 \qquad \dot{Q}_4 = (\delta/R_4)Y_2$$
$$\dot{Q}_5 = \dot{Q}_6 \qquad \dot{Q}_5 = (\delta/R_5)Y_2$$
$$\dot{Q}_6 = (\delta/R_6)Y_3$$

(a)

$$
\begin{array}{rcl}
(\delta/R_2)Y_1 \quad -\dot{Q}_2 & = & 0 \\
(\delta/R_4)Y_2 \quad\quad -\dot{Q}_4 & = & 0 \\
(\delta/R_6)Y_3 \quad\quad -\dot{Q}_6 & = & 0 \\
(\delta/R_5)Y_2 \quad\quad\quad -\dot{Q}_5 & = & 0 \\
\dot{Q}_2 & = & \dot{Q}_1 \\
-\dot{Q}_2 \; +\dot{Q}_4 \; +\dot{Q}_5 & = & \dot{Q}_3 \\
\dot{Q}_5 \; -\dot{Q}_6 & = & 0
\end{array}
$$

(b)

$$\dot{Q}_6 = [R_4/(R_4 + R_5)]\dot{Q}_1 + [R_4/(R_4 + R_5)]\dot{Q}_3$$

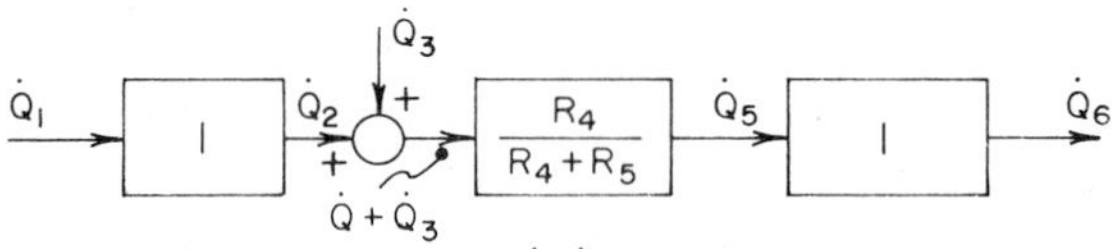

(c)

$$Y_3 = [(R_4 R_6)/\delta(R_4 + R_5)]\dot{Q}_1 + [(R_4 R_6)/\delta(R_4 + R_5)]\dot{Q}_3$$

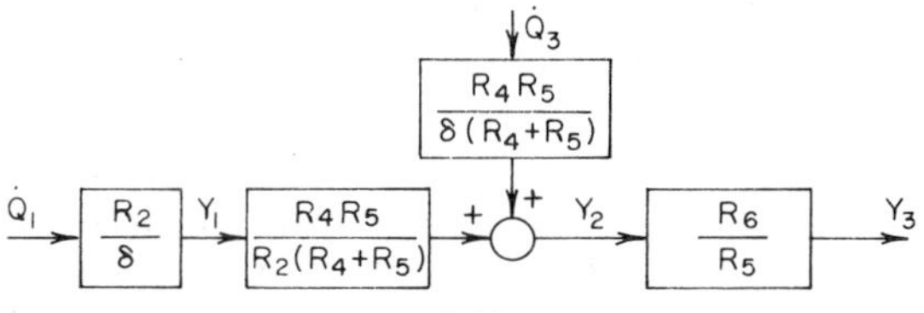

(d)

Fig. 2-15. Steady-state analysis of the three-compartment system in Fig. 2-7b. (a) Set of simultaneous equations. (b) This set of equations has been rearranged in what is termed canonical form. (c) Block diagram derived from these equations using the flow rates as dependent variables. (d) Another block diagram from the same equations in which the variables are the liquid levels.

a. The previous discussion of steady-state relations among the several variables describing a given physical system may be made more systematic in the following manner. Using the system of Fig. 2-7b as an example, the conservation law and loss functions yield the set of equations shown in (a).

b. This set of simultaneous equations may be rearranged as in (b) so as to place the terms containing the seven dependent variables on the left-hand side in columns, one for each variable. The terms appearing on the right-hand side of the equations are either independent variables or constants. One can regard $\dot{Q}_1$ and $\dot{Q}_3$ as independent variables, or inputs, either or both of which might be changed in a given problem. This set of equations is in *canonical,* or standard, form.

c. Note that there are seven equations in seven dependent variables, so that any six of the variables can be eliminated to leave a single equation in one dependent variable. Presumably $\dot{Q}_6$ is the variable of interest, and the resulting equation would be a relation between $\dot{Q}_6$, the independent variables $\dot{Q}_1$ and $\dot{Q}_3$, and the parameters. Elimination of the six variables may be accomplished by algebraic methods, or application of an algorithm such as Cramer's rule (Guillemin, 1953).

d. The physical system may be depicted by a number of block diagrams, that is, the block diagram may show varying amounts of detail ranging from a complete picture of all the subsidiary processes on the one hand, to a single block for the entire system on the other. In between these two extremes one may draw block diagrams to depict certain dependent variables and eliminate others as shown in (c) and (d).

e. The blocks appearing in a given diagram may be viewed as a physical system by which the input and output variables are related. These relations may be derived from the underlying physical laws, or in some instances by purely empirical observations. In contrast with this physical viewpoint, one may also regard the block diagram in a purely mathematical sense, in which case the functions appearing in each block are transformations of one variable into another. The algebraic expressions are the rules by which one can calculate the output variable, having been given the input.

f. When the equations are solved for a specific variable, say $\dot{Q}_6$, an expression of the form $\dot{Q}_6 = a\dot{Q}_1 + b\dot{Q}_3$ is obtained in which a and b are expressions involving the system parameters. Although in this system, a and b will both be positive, this is not always true and one or both may well be negative.

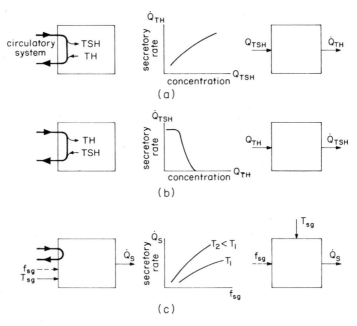

Fig. 2-16. Glands as transducers. (a) Thyroid gland which secretes thyroxine (TH) into the blood stream at a rate that depends upon the concentration of thyrotropin (TSH) in the blood. (b) Pituitary gland secretes TSH at a rate inversely proportional to the TH concentration. (c) Sweat gland secretes sweat at a rate that is a function of both a neural signal f_{sg}, and the local temperature of the gland T_{sg}.

a. The term *transducer* was introduced in Section 2.12. The glands shown in this figure are additional examples; the thyroid and pituitary glands secrete specific hormones into the circulatory system, and the sweat gland secretes sweat, the evaporation of which helps cool the body. These examples illustrate some of the relations encountered with this class of effector unit. In addition, they will serve to introduce the concept of biological fluxes, that is, other flow processes than the fluid flow contained in the previous models.

b. The thyroid gland (a) secretes thyroxine (TH) into the blood stream at a rate that is a function of the concentration of another hormone, thyrotropin (TSH), in the blood. This is suggested by the flows shown within the block. The normal concentration of TSH in the blood is of the order of 10^{-5} parts per million, so the blood simply acts as a carrier mechanism that distributes the hormone throughout the body.

c. The pituitary gland secretes TSH into the blood stream, but the secretory rate $\dot{Q}_{TSH}$ is inhibited by the concentration of TH in blood. The characteristic curve thus has a negative slope, and with $\dot{Q}_{TH}$ sufficiently large, the magnitude of $\dot{Q}_{TSH}$ can be essentially zero. The interaction between the thyroid and pituitary glands is described in Section 4.14.

d. The sweat gland (c) releases sweat to the skin surface at a rate $\dot{Q}_s$ that depends, first, upon the neural signal whose frequency is denoted f_{sg}. Experiment shows that there is an additional effect; namely, that the sweat rate $\dot{Q}_s$ also depends upon the local skin temperature, T_{sg}. The sweat rate $\dot{Q}_s$ is thus a function of two variables, and the steady-state transfer function should be represented by a surface which the two curves only suggest.

e. These three glands not only depict a variety of transducers, but also suggest three kinds of biological fluxes (Rich, 1969). A *flux of material* denotes the actual movements of molecules or other particles from one point to another under the action of a concentration gradient (diffusion) or some other transport process. Such a flux occurs in the sweat gland, that is, the water required for sweat formation must be brought to the gland by the circulatory system. The movement of hormones through the circulatory system is also a flux of material, and each of these is measured in suitable units of mass, or volume, per unit time. The transformation of one molecule to another in a biochemical reaction would also constitute a form of material flux.

f. *Flux of energy* refers to the transport or transmission of energy (thermal, chemical, electrical) over appropriate channels. In the sweat gland, its own temperature T_{sg} will be the consequence of the heat fluxes flowing into the gland and those leaving by all paths, including that of evaporation. The transport of chemical energy (as in ATP) may require an associated material flux. On the other hand, the flux of thermal energy does not require a material flux in the case of thermal conduction through tissue, but when thermal energy is transported by the blood, the quantity of energy carried is dependent upon the blood flow rate. Thermal energy flux is measured in calories or ergs per unit time.

g. A *flux of information* is found in the nervous system, in which the pulse signals are not associated with a material flux nor with an appreciable flow of energy. Hormonal signals, while they are associated with a flow of hormones, act more like a flux of information than of material.

h. Despite the convenience of the three categories from a descriptive standpoint, these fluxes cannot always be isolated but appear in various combinations, as some of the above examples suggest. Possibly of more use is the generic term *signal* to denote these fluxes as well as other physical quantities required to describe process behavior. The term signal is practically synonymous with the mathematical term variable, both denoting quantities that describe the state of the system, and which may change as the process proceeds. The term signal is widely accepted in engineering terminology, and has the advantage of suggesting the dynamic and controlling character of the physical quantity with which it is associated.

i. The distinction and interrelations between the flux of energy and material on the one hand, and that of information on the other, is discussed by Waterman (1968) and the references there cited. In most engineering texts on feedback control it is the signal transmission that is treated explicitly and the actual energy flow appears only implicity. That is, variables such as temperature, pressure, concentration, voltage, flow rate, displacement and velocity are selected as the "signals" to describe the process, with no reference to energy or power. This choice is made because equations written in terms of energy are almost always more complex, and any desired analysis more difficult to carry out.

Epitome 2.18

1. Most of the concepts associated with feedback regulating systems can be developed using the flow of fluid between tanks as an example. A steady state is said to exist when all flow rates, quantities stored, and related variables are constant.

2.   A system of flow processes in the steady state exhibits a concentration or mass gradient between compartments just sufficient to produce a mass flow rate at each point equal to the net input preceding that point.

3.   A compartment is defined as a region in which the material concentration is uniform, and for which the inlet and outlet flows can be identified. Associated with each compartment are two mathematical relations termed the storage and loss functions.

4.   The storage function relates the quantity stored to the parameters of the compartment, and the input and output flows.

5.   The loss function relates the outflow (loss) to the quantity stored, or some other variable. In many instances the loss function is derived from the law of mass action.

6.   The steady state in a given flow process is directly related to the inflow rate, which determines the quantity of material stored in each compartment.

7.   A change in any of the system parameters will in most instances result in a new steady state, but because such a change also results in a new system, comparisons between the steady states may not be too meaningful.

8.   Single as well as multicompartment systems may contain a number of inputs and outputs. In the steady state, the conservation of mass requires the equality of inflow and outflow rates, both for the entire system and for any designated portion of it.

9.   Multicompartment systems may contain processes which are coupled in either a noninteracting or an interacting manner. A noninteracting coupling is one in which the flow between two compartments depends only upon the conditions in the preceding compartment, and not the succeeding one. In contrast, the flow between interacting compartments depends upon the conditions in both.

10.   The flow of aqueous through the eye may be represented by a two-compartment model. Changes from the normal steady state may result in excessive pressures within the eye and glaucoma.

11.   Disturbances to the steady state of a flow system may take the form of a change in the input rate, or a change in parameters. Analogous systems are defined as those represented by equations of the same form.

12.   The steady-state transmission (or gain) of a flow process is the ratio of a change in the output flow rate to the change in an input flow rate producing it. For a single input, single output flow process, this ratio is unity, but with additional inputs and outputs the gain may be other than

2.18   EPITOME   41

one. More generally, the input and output may be unlike physical quantities, and the component is then termed a transducer.

13. Flow processes may be represented by flow graphs which depict the material flow between compartments, and are based upon the conservation of mass. Alternatively, block diagrams are drawn to show the functional relations between pairs of signals (input, output) associated with each subsidiary process. The block diagram is not unique, but may be drawn to emphasize those signals of interest in a given problem. Thus it may be a substantial abstraction of the physical system.

14. Intravenous feeding provides an example of a three-compartment, interacting system. The net flow of glucose between compartments depends upon the difference in glucose concentrations, and a steady state may be attained only if the feeding rate is less than the maximum metabolic rate for glucose in the cells.

15. The block diagram mode of analysis is applicable to many systems other than the compartmental systems associated with material flow. The neuromuscular control of eye position may be represented by block diagrams in which each block has an output related to its input in an appropriate functional manner.

16. A quantitative analysis of the steady state requires that the equations defining the relations between all pertinent variables are written and then solved for the quantities of interest. A number of block diagrams may be constructed for this set of equations; the form of the diagram and content of the blocks will vary greatly, depending upon which set of variables is used.

17. Glands form a class of transducers in which the secretion rate (output) is some function of an input variable. For descriptive purposes, it is sometimes helpful to speak of a flux of material, energy, or information, but these three fluxes are often interrelated and cannot be isolated. Hence the more general term signal is applied to all biological variables, and the analysis problem is one of relating the signals to each other in an appropriate functional manner.

Problems

1. Application of the term steady state requires a clear definition of the process or system to which it refers.

(a) Consider a flashlight: battery, wiring, switch, and lamp. In what sense can it be said that a steady state exists when the light is on? In what sense is this not a steady state?

2   FLOW PROCESSES IN THE STEADY STATE

(b)   In the study of work performed by humans, the work rate may be held at a constant value. Can this be considered a steady state? Discuss the limitations of this viewpoint.

2.   With reference to Fig. 2-2, under what physiological conditions will (a) $\dot{Q}_a > \dot{Q}_v$, or (b) $\dot{Q}_a < \dot{Q}_v$?

3.   Hydraulic flow resistance may be defined by the ratio $\delta Y/\dot{Q}$. Show that R has the dimensions given in Fig. 2-3.

4.   The quantity $\dot{Q}$ is the time rate of flow (of material or energy) across some surface. Identify the surfaces for $\dot{Q}_3$ and $\dot{Q}_2$ in Fig. 2-1a, and for $\dot{Q}_v$ and $\dot{Q}_2$ in Fig. 2-2a.

5.   Consider a single tank, as in Fig. 2-3, but having an additional input $\dot{Q}_2$ whose value is constant at all times. Using physical reasoning, plot $\dot{Q}_0$ as a function of Y. Does R remain the same as in Fig. 2-3?

6.   Indicate the consequences of the following disturbances in Fig. 2-9.

Disturbances

R_1	R_2	A_1	A_2	$\dot{Q}_1$	$\dot{Q}_4$
$+\Delta$	$-\Delta$	0	0	0	0
$+\Delta$	0	0	0	$-\Delta$	0
$+\Delta$	0	0	0	$+\Delta$	$+\Delta$

7.   Confirm the equations and block diagrams in Fig. 2-14. These pertain to the system shown in Fig. 2-7b.

8.   Write the set of equations for Fig. 2-8c and place them in canonical form.

9.   Consider the three interacting compartments shown in Fig. P2-9.

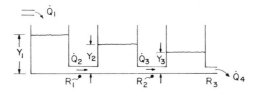

Fig. P2-9

(a)   If $R_1 = R_2 = R_3$, show that $Y_1 = 3Y_3$, and $Y_2 = 2Y_3$.
(b)   If $R_1 = 2R_2 = 3R_3$, find the relative values of Y_1, Y_2 and Y_3.
(c)   If $(Y_1 - Y_2) = 3(Y_2 - Y_3)$, what is the relative value of R_1 and R_2? What will R_3 be in terms of the other quantities?

10. Three noninteracting processes are coupled together as shown in the block diagram of Fig. P2-10 and have the following steady state relations.

$$X_2 = a_1X_1 + b_1,$$
$$X_3 = -a_2X_2 + b_2,$$
$$X_4 = a_3X_3 - b_3.$$

(a) Sketch the steady-state characteristic curve for each block in the diagram.

(b) Calculate the steady-state value of X_4 for a given value of X_1.

(c) Calculate the incremental gain $\Delta X_4/\Delta X_1$.

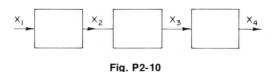

Fig. P2-10

11. (a) Write the steady-state equations for the system of noninteracting compartments in Fig. P2-11a, and derive the steady-state transfer function $\dot{Q}_3/\dot{Q}_1$.

(b) Write the equations and derive the transfer function as in (a) for the same two compartments connected in an interacting manner, as shown in Fig. P2-11b.

(c) Will the two systems store the same amount of material if the inputs are identical? Explain.

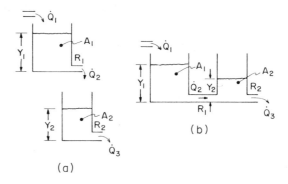

Fig. P2-11

12. Confirm the equations given in Fig. 2-11.

References

Becker, B. and Shaffer, R. N. (1961). "Diagnosis and Therapy of the Glaucomas." C. V. Mosby Co., St. Louis.

Brownell, G. B. and Ashare, A. B. (1967). Definition of pool and space. AEC Conf. 661010, *Compartments, Pools, and Spaces.*

Guillemin, E. A. (1953). "Introductory Circuit Analysis." Chap. 3, Wiley, New York.

Rich, A. (1969). Review of "Comprehensive Biochemistry." *Sci. Amer.* **220,** Feb., 126.

Riggs, D. S. (1963). "Mathematical Approach to Physiological Problems." Chapter 7. Williams & Wilkins, Baltimore, Maryland.

Riggs, D. S. (1970). "Control Theory and Physiological Feedback Mechanisms." Williams & Wilkins, Baltimore, Maryland.

Waterman, T. H. (1968). Systems theory and biology — View of a biologist. *In* "Systems Theory and Biology," (M. D. Mesarovic, ed.), pp. 1–37. Springer-Verlag, Berlin and New York.

Dynamic Behavior; the Transient Response || *Chapter 3*

If one thing in nature is constant, it is the fact of continual change. This truism is nowhere better demonstrated than in the biological processes that constitute life, and it is against this background of change that we propose to discuss homeostasis.

In Chapter 2, the concept of the steady state was developed, although it was made clear that a physical steady state is rarely attained by an organism except possibly in sleep or controlled experimental conditions. A large portion of an animal's life is spent in moving from one steady state to another, without reaching the second before a new disturbance appears.

The examples previously described were almost exclusively linear systems, a fact that has two implications. The steady-state characteristic relations are straight lines, so that a change in one variable brings about a strictly proportional change in a related dependent variable. Second, it is possible to superimpose two input changes and ascertain the change in the output by adding the changes produced by the two input signals when observed separately. This *principle of superposition* is an extremely valuable tool and is widely applied when it is clearly demonstrated that the system is linear and the application of the principle is thus valid. Conversely, if superposition is shown to be valid experimentally, this is a suitable test for establishing linearity.

In the present chapter, the question of change will be considered for linear systems with the objectives of understanding the changes which biological systems undergo following a disturbance and of developing a vocabulary that will permit a precise description of such changes. Our interest in the dynamics of biological systems stems from the salient role that this plays in regulatory behavior, as well as from the fact that dynamic behavior reveals more about the processes than can possibly be obtained from steady-state observations alone.

The following discussions will make extended use of the superposition principle. A disturbance to a linear system usually produces a transitory dynamical response followed by a continuous change in the steady state that continues as long as the disturbance persists. The transient portion represents the adjustments within the system necessary for it to conform to the disturbance. The transient and steady-state components can be calculated separately, and the total response can be found by adding them. The superposition principle permits the analysis of complex behavior in terms of relatively simple components, a procedure not possible with a system containing nonlinearities.

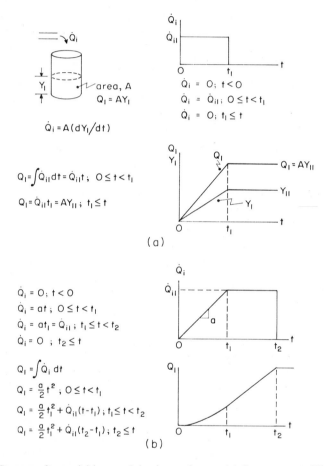

Fig. 3-1. Storage of material in a tank having no losses. (a) Compartment filling at a constant rate, and (b) filling at a linearly increasing rate. The tank accumulates the inflow so that the quantity stored is the integral over time of the past inflow.

3   DYNAMIC BEHAVIOR; THE TRANSIENT RESPONSE

a. We now turn to the manner in which a given process moves from one steady state to another, and specifically, the situation in which the hydrostatic level Y_1, the quantity stored Q, and the flow rate $\dot{Q}$ are changing. This, termed the *dynamic behavior,* consists of transient changes that take place as a process adjusts itself to a new set of conditions, that is, as it goes about reaching a new steady state.

b. Consider a tank as in (a), into which liquid flows at the rate $\dot{Q}_1$, but from which there is no loss of liquid, that is, no outflow or loss function. This is an example of pure storage. If one imagines that the tank is initially empty ($Y_1 = 0$), and that $\dot{Q}_i$ is then established at a constant magnitude for t_1 seconds, the material accumulates at a constant rate and the level Y_1 rises linearly with time. This continues until $\dot{Q}_i$ falls to zero and Y_1 remains at its maximum value Y_{11}.

c. The constant inflow rate in this example is purely arbitrary, and was selected to show that the tank accumulates the inflow. In mathematical language, the volume stored at any instant (Q_1) is the integral over time of past inflows. Thus $Q_1 = \int \dot{Q}_i\, dt$, and when $\dot{Q}_i$ falls to zero at t_1 there is no further contribution to the integral, so that Q_1 remains constant at its maximum value $Q_1 = \int_0^{t_1} \dot{Q}_i\, dt$.

d. To carry out the above integration, one must express $\dot{Q}_i$ as a function of time. Thus in (a), during the interval 0 to t_1, $\dot{Q}_i$ is a constant ($\dot{Q}_{i1}$) and $Q_1 = \int \dot{Q}_{i1}\, dt = \dot{Q}_{i1}t$. In (b), over the interval 0 to t_1, $\dot{Q}_i = at$, where a is a constant, and the integral becomes $Q_1 = at^2/2$ as shown. For the subsequent time interval t_1 to t_2, the volume stored, Q_1, increases linearly with time since $\dot{Q}_i$ is a constant, $\dot{Q}_{i1}$.

e. A physical interpretation of this figure may be developed by noting that when $\dot{Q}_i$ is a constant (interval t_1 to t_2), the volume stored, Q_1, increases in a linear fashion. On the other hand, when $\dot{Q}_i$ increases in a linear fashion (0 to t_1), Q_1 increases more rapidly as time goes on, as given by the curve $Q_1 = at^2/2$.

f. This figure shows no means of removing material from the tank and so the integral $\int \dot{Q}_i\, dt$ can only increase with time. If, however, a pump or or other means is provided to permit removal of liquid at an arbitrary rate $\dot{Q}_0$, then Q_1 can either increase or decrease with time. $\dot{Q}_0$ can be considered a negative inflow, so that with both inflow and outflow the integral becomes $Q_1 = \int (\dot{Q}_i - \dot{Q}_0)\, dt = \int \dot{Q}_i\, dt - \int \dot{Q}_0\, dt$.

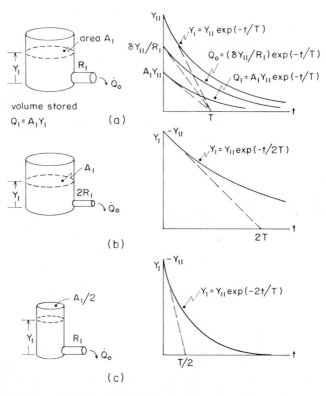

Fig. 3-2. The effect of compartment size and outflow resistance on the loss of material from a compartment having no inflow. (a) Exponential decay upon emptying, with a time constant T. (b) Same, with time constant $2T$. (c) Same, with time constant $T/2$.

3   DYNAMIC BEHAVIOR; THE TRANSIENT RESPONSE

a. The single tank or compartment has no inflow and we are only concerned with the manner in which it empties through a resistance R_1. The outflow is assumed to be proportional to Y_1 at each instant, or to the volume stored, $A_1 Y_1$.

b. The three curves in (a) show the manner in which the quantities Y_1, $\dot{Q}_0$, and Q_1 all change with time. Inasmuch as these variables are all proportional to each other ($Q_1 = A_1 Y_1$; $\dot{Q}_0 = \delta Y_1 / R_1$), the curves are identical except for vertical scales.

c. These curves represent the *exponential function* $\epsilon^{-t/T}$, where ϵ is the constant $2.7318\ldots$, t is time, and T the *time constant,* as described in the next section. For typographical convenience, the exponential terms will be written as $\exp(-t/T)$, as in the figure.

d. The exponential curve is characterized by an initial rate of change that is large, but which decreases with time so that the rate of change, or slope, becomes very small as $t \to \infty$. Viewed physically, the tank empties rapidly at first because of the high hydrostatic pressure, but as the tank empties the variables $\dot{Q}_0$ and Y_1 decrease more slowly and eventually approach zero. Further properties of the exponential curve appear in later sections.

e. Factors which affect the rate of emptying are shown in (b) and (c). An increase in the outflow resistance to $2R_1$ doubles the time constant, and the tank takes twice as long to empty. On the other hand, a decrease in size of the tank to $A_1/2$ results in the tank emptying more rapidly. However, the two curves (b) and (c) are similar, and would superimpose upon a suitable change in the time scale. Curve (b) has a time constant of $2T$, and in (c) the time constant is $T/2$.

f. The exponential curves in this figure approach their final values (zero) with increasing time. Mathematically, the exponential $\exp(-t/T)$ approaches the value zero as t becomes very large. However, the curve is asymptotic to the time axis and never quite reaches zero for any finite time. From a practical standpoint the physical quantity represented by the exponential approaches so close to zero that the difference cannot be measured.

g. The exponentially decaying transient, also termed a *first-order transient,* is of considerable general interest. In more complex systems with many interacting compartments, the transient curves will not be of this simple form, but can frequently be described as the sum of a number of exponential components. This holds true only for linear systems, but in such cases the exponential transient becomes a fundamental building block in the discussion of dynamic behavior.

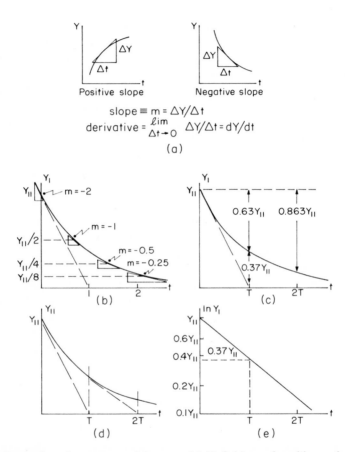

Fig. 3-3. Properties of an exponential curve. (a) Definitions of positive and negative slopes. (b) Change in the slope of an exponential curve with the decay in its magnitude. (c) Decay in magnitude in an interval of one time constant. (d) Construction of an exponential curve using slopes. (e) Semilogarithmic plot.

a. The exponential curve has a number of distinctive properties which are helpful in interpreting its physical implications as well as its geometric shape (Riggs, 1963).

b. A property of any curve is its *slope* (*m*) which is defined in (a). The increments ΔY and Δt are given by the sides of a triangle in which the hypotenuse is drawn tangent to the curve at a given point. The slope may be either positive or negative (Daniels, 1956).

c. The ratio $\Delta Y/\Delta t$ leads to the mathematical concept of the *derivative,* which is the slope of the curve at a specific point. Generally, a curve will have a derivative (slope) that is different at each point along the curve, whereas a straight line has the same derivative everywhere.

d. The derivative of the exponentially decaying curve is large and negative at $t = 0$, becomes less as t increases, approaching zero as t becomes infinite. Curve (b) has been arbitrarily drawn with $m = dY_1/dt = -2$ at $t = 0$. When $Y_1 = Y_{11}/2$, $m = -1$, and so on. Note that when Y_1 is reduced by half, dY_1/dt is also reduced by the same amount. In general, the slope of this decaying exponential curve is proportional to the negative of the magnitude of Y_1 at each point; this is a fundamental property of the exponential curve.

e. In curve (c), the initial slope has been projected down to the time axis, which it intersects at $t = T$. T is termed the *time constant,* and is one way of characterizing the curve. Detailed calculation will show that at $t = T$, the magnitude of Y_1 has fallen to $0.37Y_{11}$. A similar relation holds for the next T seconds, after which Y_1 has reached $(0.37 \cdot 0.37)Y_{11} = 0.137Y_{11}$. This curve, starting at any value of Y_1, will decay by 63% during the next T seconds.

f. The slope at any point of the curve may be obtained by drawing the tangents as in (d).

g. If one plots log Y_1 (rather than Y_1) as a function of time, the curve becomes a straight line as in (e); this is termed a semilogarithmic plot. Note that the vertical axis has a logarithmic scale. Thus an empty tank ($Y_1 = 0$) is not shown, although the vertical scale could be extended to cover as many decades as desired.

h. The exponential decay curve occurs whenever the rate of change of a variable (dY_1/dt) is proportional to the negative of its magnitude, that is, $dY_1/dt = -aY_1$. The constant a turns out to be $1/T$. This differential equation, and equations of a similar form, thus have as solutions a decaying exponential, such as $Y_{11}\epsilon^{-t/T}$. The exponential function is a fundamental mode of change for physical systems described by such differential equations.

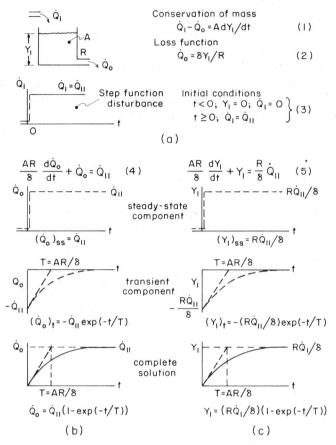

Fig. 3-4. Dynamic response of a first-order process to a step function disturbance. (a) Process, disturbance, and initial conditions. (b) Steady-state and transient components together with the complete solution for $\dot{Q}_0$. (c) Same for Y_1.

3   DYNAMIC BEHAVIOR; THE TRANSIENT RESPONSE

a. The compartment (tank) having both inflow ($\dot{Q}_1$) and outflow ($\dot{Q}_0$) is described by Eqs. (1) and (2) in the figure; these are the same storage and loss functions described previously. If the inflow is suddenly changed, say from zero to some constant value $\dot{Q}_{11}$, the disturbance is termed a *step function*. If the tank was initially empty, the initial conditions and the disturbance are described by Eqs. (3).

b. Equations (1) and (2) contain two dependent variables, $\dot{Q}_0$ and Y_1, either one of which may be eliminated to obtain a differential equation in terms of a single variable. Equation (4) is the result of eliminating Y_1, and the complete solution along with its components is shown in (b).

c. The solution of this ordinary linear differential equation contains a *steady-state* component plus one transient component. Physical reasoning suggests that as $t \to \infty$, $\dot{Q}_0$ ceases to change, so that in the steady-state $d\dot{Q}_0/dt = 0$, and $(\dot{Q}_0)_{ss} = \dot{Q}_{11}$.

d. The transient component $(\dot{Q}_0)_t$ is exponential and has the magnitude and sign required for the complete solution to match the physical conditions at $t = 0$, that is, $(\dot{Q}_0)_{ss} + (\dot{Q}_0)_t = 0$. It likewise satisfies the physical conditions for large values of t, that is, $(\dot{Q}_0)_{ss} + (\dot{Q}_0)_t = \dot{Q}_{11}$.

e. In this single-compartment (first-order) process there is only one transient term, but in general there will be as many transient components as there are compartments. In any case, the transient components describe the way in which a system moves from one steady state to another. If the initial conditions or disturbance had been different, the complete solution, or response curve, would also be altered in a manner so that it would still yield the correct initial and final conditions.

f. The differential equation and complete solution for Y_1 are shown in (c). Note that the transient components in the two solutions are similar, and that the time constants, which are characteristic of the system, are identical.

g. From a physical standpoint, $\dot{Q}_0$ and Y_1 increase as the tank fills up, but the rate at which they increase depends upon the difference $\dot{Q}_1 - \dot{Q}_0$. This difference is a maximum at first, but then decreases to zero as the tank fills up.

h. In finding the complete solution to a dynamic problem, there will be n constants of integration to evaluate, where n is the order of the equation. These are computed from n initial conditions, and provide the means by which the general solution is made to fit the precise conditions of a specific problem (Trimmer, 1950).

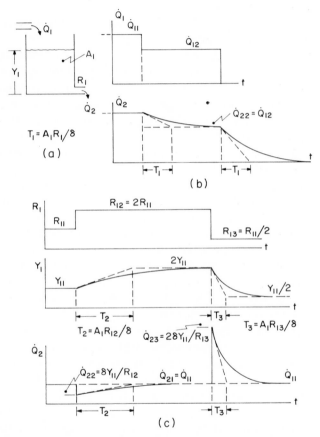

Fig. 3-5. First-order responses to a sequence of step function disturbances. (a) First-order fluid flow process. (b) Behavior of $\dot{Q}_2$ following step function decreases in $\dot{Q}_1$. (c) Behavior of $\dot{Q}_2$ following step function changes in R_1.

   a. Typical disturbances to the first-order process in (a) might be a change in the inflow, $\dot{Q}_1$, or a change in R_1. In (b) the inflow is reduced in two separate steps, each of which results in an exponential transient in $\dot{Q}_2$. and ultimately in a new steady state. The process time constant remains the same for both disturbances inasmuch as $\dot{Q}_1$ does not affect $T = A_1 R_1/\delta$. Thus both exponential transients are basically the same, only with different magnitudes.

   b. The effect of a change in the outlet resistance is shown in (c). The dynamic character of the responses differ in two significant ways from the previous case. In the first place, since $\dot{Q}_1$ is not changed, the steady-state value of $\dot{Q}_2$ remains unaffected, even though there are transient changes in the outflow. Second, a change in R_1 changes the time constant, resulting in different values of T for the increase and decrease in R_1. Note also that although there is no change in the steady-state value of $\dot{Q}_2$, there is a change in Y_1 in the steady state.

   c. The sudden changes in $\dot{Q}_2$ that follow the increase and decrease in R_1 are the consequence of the implicit assumption that no time is required to accelerate the fluid flow following the resistance change. Stated somewhat more precisely, this implies that the accelerating time for the fluid is very short compared to the tank time constant. On the other hand, discontinuous changes in Y_1 are not possible because this would be tantamount to moving a finite quantity of material in zero time.

   d. The previous discussion has shown that despite the fact that the disturbances to this process can be quite varied and may appear at a number of points in the system, the transient behavior is of an exponential character with a time constant corresponding to the process parameters *after* the disturbance has appeared. Furthermore, the same exponential component appears in the response of all the process variables. However, the magnitude of the transient term varies for the different variables; it is always of such a magnitude as to effect a transition from the initial conditions to the final steady state.

   e. In both of the above cases, the disturbance was assumed to be a step function change. Other forms for the disturbance can be readily visualized, but in all cases the transient component will be an exponential term having an appropriate time constant. The choice of a step function in this discussion was dictated solely by simplicity of exposition.

3.6   FIRST-ORDER DYNAMICS

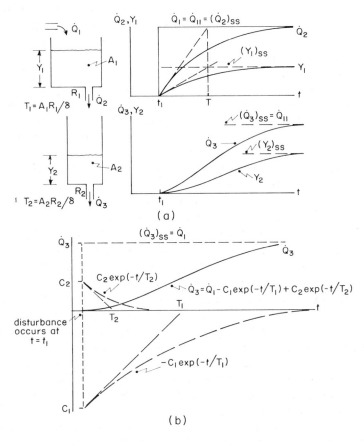

Fig. 3-6. Responses of a second-order system of noninteracting processes to step function changes in the input. (a) First- and second-order responses of the dependent variables. (b) Resolution of a second-order response into its steady-state and transient components.

a. Two first-order processes, connected so that the efflux of one becomes the influx of the second, form a second-order system. The concept
of system order is associated with the number of coupled or interconnect-
ing compartments, each of which contributes to the dynamic behavior of
the output when the system is disturbed. An n-compartment system of
this character is an nth-order system.

b. The two compartments in this figure are coupled in a noninteracting
manner and therefore a step function change in $\dot{Q}_1$ will cause Y_1 and $\dot{Q}_2$ to
change in an exponential manner. When this occurs, the influx to the
second compartment $\dot{Q}_2$ is exponential (not a step function) and the
variables Y_2 and $\dot{Q}_3$ exhibit a sigmoid form. $\dot{Q}_3$ lags behind $\dot{Q}_2$ for the same
reason that $\dot{Q}_2$ lags behind $\dot{Q}_1$, and $\dot{Q}_3$ consists of the sum of two exponentials.

c. The response of $\dot{Q}_3$ to this disturbance may be shown to contain
three components.
  (1) The steady state, $\dot{Q}_3 = \dot{Q}_1$.
  (2) A transient, $-C_1 \exp(-t/T_1)$, contributed by the first tank.
  (3) A transient, $+C_2 \exp(-t/T_2)$, contributed by the second tank.
The constants $-C_1$ and C_2 have magnitudes such that the sum of the three
components is zero at $t = 0$. The transients disappear as $t \to \infty$.

d. The above discussion may be generalized as follows. An nth-order
system containing n interconnected compartments will have a transient
response consisting of n exponential terms, providing that the disturbance
and the response are at such points in the system that all compartments
participate.

e. If the compartments are coupled in a noninteracting manner (as in
this figure), then the time constants in the transient response are identical
with the time constants of the individual compartments. This is not true
for systems of interacting compartments.

f. In the initial conditions prior to the disturbance, the values of all
system variables, including the dependent variables, were assumed to be
zero. This fact was used to compute the constants C_1 and C_2. More
generally, the initial values may be other than zero, and the constants,
whose magnitudes depend upon both the initial conditions and the disturbance, may be either positive or negative. Furthermore, for various
disturbances and initial conditions, the complete response may appear to
be quite different from that shown in the figure. In all cases, however,
the same exponential terms appear, although combined with different
magnitudes and signs.

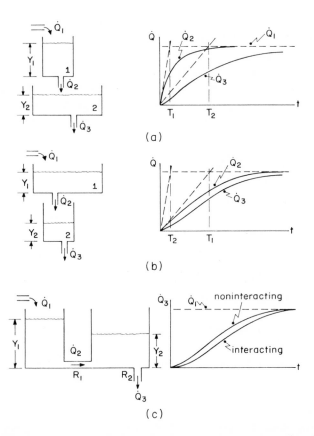

Fig. 3-7. Responses of several different second-order systems to step function disturbances at the input. (a) Two noninteracting compartments. (b) Same, but with the two compartments interchanged. (c) Two interacting compartments.

3   DYNAMIC BEHAVIOR; THE TRANSIENT RESPONSE

a. A somewhat curious relation arises from a consideration of the sequence in which processes are arranged within the system. Parts (a) and (b) of the figure show two dissimilar and noninteracting capacities arranged in the two possible sequences. In (a), the smaller compartment (smaller time constant) is first, and $\dot{Q}_2[t]$ is appreciably more rapid in response than $\dot{Q}_3[t]$. The latter, $\dot{Q}_3[t]$, exhibits the sigmoid character discussed previously.

b. If the sequence is reversed (and the tanks renumbered), the first tank has the larger time constant, but the final responses of both systems, $\dot{Q}_3[t]$, are identical. It is characteristic of noninteracting systems that the sequence of component processes is immaterial as regards overall stimulus–response relationships. This fact holds only for linear systems in which the storage and loss functions obey the same relations as previously described.

c. The previous examples have been restricted to the noninteracting cases simply because they are more readily understood, and physical intuition can help in analyzing them. However, when the processes "interact" as in (c), such that changes in Y_1 depend not only on conditions in the first compartment but on those in the second tank as well, it becomes practically impossible to use only physical intuition. $\dot{Q}_3$ might be expected to exhibit a sigmoid character following a step change in $\dot{Q}_1$, but the difference in $\dot{Q}_3$ between the noninteracting and interacting cases is not obvious.

d. The similarity in responses for the two cases resides in the fact that they are both sigmoid, and each can be resolved into a steady state and two exponential transient components, as in Fig. 3-6b.

e. The distinction between the two responses lies in the time constants characterizing the transients. Whereas in the noninteracting case each time constant was uniquely determined by the parameters of one compartment, $T_1 = A_1 R_1/\delta$, in the interacting case the two time constants involve parameters from *both* processes. That is, although both transients are made up of two exponential components, the magnitude of the time constants may be quite different, and in the interacting case under consideration both time constants are calculated from parameters of both processes. Unfortunately there is no simple way of inferring this, or of calculating their values.

f. A second distinction between the two $\dot{Q}_3$ curves lies in the fact that $\dot{Q}_3$ for the interacting case always lags behind its counterpart for the noninteracting case, as is clearly shown in (c). For comparison it is assumed that the noninteracting case is that obtained by separating the two tanks, each with the same area and resistance but arranged as in (a) or (b).

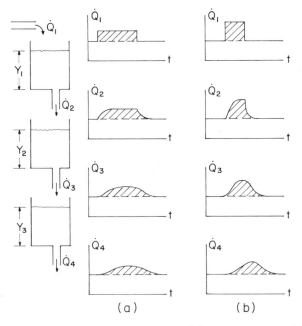

Fig. 3-8. Pulse transmission through a sequence of three compartments. (a) Long pulse of low amplitude, and (b) shorter pulse of equal area. The transmission is characterized by (1) decrease in maximum amplitude, (2) increase in duration, and (3) conservation of area. The latter is distinctly a special case due in this instance to the conservation of mass.

a. In a physical system of many compartments, the manner in which a disturbance is transmitted from one process to the next is a principal concern. This transmission of a disturbance may be termed *signal transmission,* the word signal being used in the very general sense described in Chapter 2. An important question involves the fidelity with which a signal is transmitted through a sequence of compartments.

b. Three noninteracting compartments are shown in this figure. Assume first that the input $\dot{Q}_1$ is changed in a step-like manner and after a short interval returned to its initial value as in (a). $\dot{Q}_2$ has the shape one would expect for a first-order process, that is, single exponential transients. $\dot{Q}_3$ has the sigmoid character typical of a second-order system. $\dot{Q}_4$ has a similar sigmoid shape, but is actually composed of three exponential transients.

c. Inasmuch as the output from each of the compartments is somewhat delayed compared to the input signal to that compartment, it may well be that $\dot{Q}_4$ never reaches a steady-state value before it starts to decline. This is possibly shown more dramatically in (b), where the input pulse has a greater amplitude, but a shorter duration.

d. This figure illustrates the fact that because of storage and the resistance to flow in each of the several compartments, the shape of a disturbance signal is greatly modified as it is transmitted through the system. This modification of the transmitted signal takes two forms.

    (1)   The amplitude tends to be decreased so that the maximum value of $\dot{Q}_4$ is less than that of $\dot{Q}_1$.

    (2)   There is a progressive delay in the signal as it passes through successive compartments.

Both of these factors tend to become more pronounced as the duration of the disturbing signal is made small compared to the system time constants.

e. A fluid flow system, as in this figure, has an additional constraint in that mass is conserved. The shaded areas in the figure are all equal because the pulses in $\dot{Q}_1$ have been made of equal area. The fact that the areas are equal follows from the requirement that all the volume entering the system during the pulse must leave before the final steady state is attained.

f. In a more general case in which the conservation of mass does not apply, although the pulse area is not conserved, the other observations are still valid. A pulse signal is attenuated in its transmission through a number of components, and is also delayed in time. The pulse becomes distorted.

3.9 SIGNAL TRANSMISSION   63

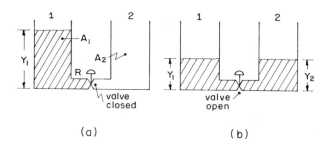

(a)   (b)

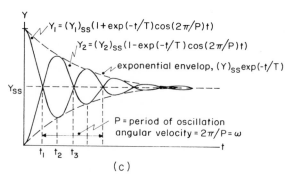

(c)

Fig. 3-9. Second-order fluid flow system with oscillatory response. (a) Initial state, (b) final state, and (c) damped oscillatory response having the general form $Y_{ss}[1 + C_1 \exp(-t/T) \sin(\omega t + \psi)]$, where C_1 and ψ are integration constants calculated to match the initial conditions, ω is the angular velocity in radians per second, and T is the time constant of the exponential envelop. Either the sin or cos function may be used with appropriate choice of ψ. See Chapter 6 for a discussion of sinusoidal signals.

a. In all the previous cases, the behavior of the system variables has been a smooth, monotonic transition from some initial conditions to a final steady state. The term *monotonic* signifies that a variable continues to increase (or decrease) during the entire time that it is changing, that is, it never reverses its direction of movement, as does the variable Y_1 in this figure. However, oscillations such as exemplified here are also a possible form for the transient response.

b. It is assumed that the liquid is initially all stored in the first tank with the valve closed. The second tank is initially empty, and for convenience is assumed to have the same volume as the first ($A_2 = A_1$). Upon opening the valve, material flows into the second tank and, because the two areas are equal, will ultimately divide itself equally between the two tanks so that $(Y_1)_{ss} = (Y_2)_{ss}$. The system is then in a steady state, and also in equilibrium because no further flow is possible.

c. During the transient period it is possible, if the conditions are right, for the variables to oscillate as shown. Y_1 and Y_2 are seen to overshoot their final values, and may do so several times before settling down to the final steady state. At $t = t_1$ there is an equal division of material between the two tanks because $Y_1 = Y_2$, but Y_2 continues to rise and Y_1 to fall until t_2, at which time the flow reverses and Y_1 starts to rise and Y_2 to fall. What accounts for this overshoot and oscillation?

d. The question of whether or not oscillations will appear is primarily one of distribution of stored energy. Initially, energy is stored in the first tank in the form of *potential energy,* as indicated by the hydrostatic head Y_1, which will cause flow into tank two when possible. The flowing liquid acquires *kinetic energy* by virtue of its velocity, and it is this "force" of the kinetic energy which enables the liquid to continue to flow into the second tank even though $Y_2 > Y_1$. Kinetic energy can cause the flow to continue even though it is against the hydrostatic head.

e. The oscillations eventually disappear; they are said to be *damped,* and the character of this damping is depicted by the dotted envelope, which, in this example and in linear systems generally, is exponential in form.

f. In other types of physical systems, damped oscillations of this same form are readily found, but the identification of the energy stores may not be at all obvious. At least two compartments are required before oscillations are possible, and even then oscillations only appear when the several parameters describing the system have appropriate values.

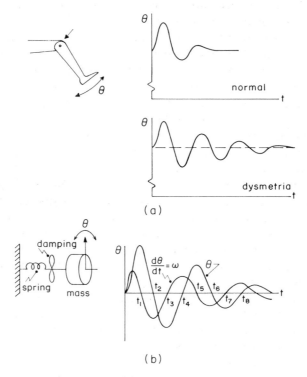

Fig. 3-10. Damped oscillations in mechanical systems. (a) Knee jerk under normal and abnormal conditions. (b) Rotational system having mass, elastic restoring torque (spring), and viscous damping.

a. In normal biological systems it is rare to encounter a prolonged oscillatory transient of the character described in the previous section. Voluntary limb motion is almost entirely "dead-beat," that is, the member reaches its final position with little or no overshoot. However, under pathological conditions, or dysfunction, transient oscillations may be pronounced.

b. *Dysmetria* is the term applied to oscillations in movement resulting from some impairment of cerebellar function and is observed in the knee jerk (Holmes, 1922; Ruch and Patton, 1965) and eye movement (Cogan, 1954). The cerebellum, in communication with the proprioceptors and the motor cortex, normally compensates for any oscillations which are inherent in the limb itself. In the normal individual the compensatory signals provided by the cerebellum greatly reduce all tendency to oscillate, and oscillations only become apparent when the compensation is lost.

c. The knee jerk provides a good example. The tendon tap is an impulse-like disturbance, which in normal man results in an oscillation in leg position that is rapidly damped. With dysmetria the oscillations may persist for 3 or 4 cycles before being damped out.

d. This behavior may be compared to that of a rotating mass attached to a torsion spring. If the mass is given a sharp twist or torque, as from a hammer blow, it will oscillate as shown, and the oscillations will be damped out as a result of air friction.

e. When the angular velocity $\omega = d\theta/dt$ is plotted along with the displacement θ, is is clear that ω is zero when θ is at its maximum (or minimum) values. Likewise ω is a maximum (minimum) when $\theta = 0$. The energy imparted to the mass by the impulse is then stored all in the spring at t_1, t_3, t_5, and t_7, that is, when $\omega = 0$.

f. Examination of this figure will show that the energy within the system is stored first all in the spring (t_1) and then, as kinetic energy, in the mass (t_2), and so on alternately. This is entirely equivalent to the process described in Fig. 3-9.

g. Transient oscillations of this or a similar character can only take place if the process is of such a type that energy can be stored (temporarily) in two different forms or locations. Furthermore, there must be a way in which energy can flow between these two forms. Dissipation of the energy as heat produced by the flow resistance accounts for the damping and eventual disappearance of the oscillations.

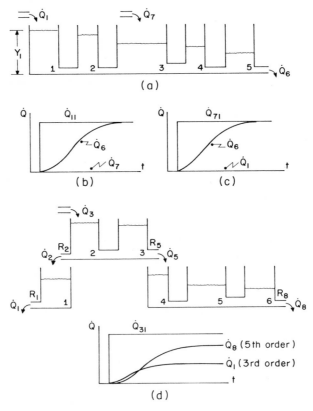

Fig. 3-11. Dynamic behavior of high-order systems. (a) Five-compartment system with (b) response to step function input at $\dot{Q}_1$, and (c) response to step function input at $\dot{Q}_7$. (d) Six-compartment system containing both interacting and noninteracting processes. It exhibits a third-order response at $\dot{Q}_1$, and a fifth-order response at $\dot{Q}_8$ to changes in the input $\dot{Q}_3$.

a. One measure of system complexity is provided by the number of coupled compartments that affect a given disturbance-response experiment, and this number may be defined as the *system order*. The determination of system order requires careful examination of the coupling between compartments.

b. There are five coupled compartments in (a), and the response of $\dot{Q}_6$ for a step function change in $\dot{Q}_1$, as sketched in (b), contains five exponential components. The order of the system, and therefore the number of transient components, is independent of the form of the disturbance. The time constants for the five exponential terms are not the same as the time constants for the individual compartments because they are coupled in an interacting (bilateral) manner (Section 2.9). They would have to be calculated from a fifth-degree polynomial describing the entire system.

c. If the disturbance had been a change in $\dot{Q}_7$, the response of $\dot{Q}_6$ would again contain five exponential terms having the same time constants as in the previous experiment, but the several terms would probably have different amplitudes than in the previous case. The response as shown in (c) looks very much the same. Note, however, that in the steady state, with $\dot{Q}_1 = 0$ and $\dot{Q}_7 \neq 0$, compartments 1, 2, and 3 reach an equilibrium (equal values of Y) and there is no flow between them. However, these compartments do participate in any transient process.

d. The six compartments in the system shown in (d) yield responses of different order depending upon the location of the disturbance and the observed response. For a disturbance to $\dot{Q}_3$, the response observed at $\dot{Q}_2$ would be of the second order and that at $\dot{Q}_1$ of the third order. On the other hand, a response to this same disturbance observed at $\dot{Q}_8$ would be of the fifth order, and contain five exponential terms. A comparison of the behavior of $\dot{Q}_1$ and $\dot{Q}_8$ in (d) shows that the fifth-order system takes longer to reach a steady state. The fact that $\dot{Q}_8 > \dot{Q}_1$ follows from the fact that $R_2 > R_5$.

e. The compartmental systems as shown in this figure are unlikely to exhibit oscillations as suggested by Fig. 3-9, simply because biological materials do not acquire sufficient kinetic energy. However, oscillation in a system of this character may well appear if the system is regulated by feedback processes as discussed subsequently in Chapter 5.

f. System order is the same as the number of coupled first-order processes, the order of the differential equation describing the system, and the number of transient terms in the complete solution.

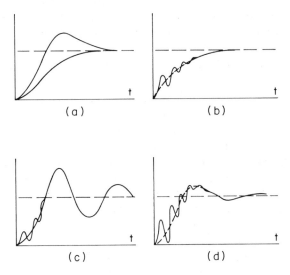

Fig. 3-12. Responses of nth-order systems for which $n = 2p + q$, where n is the system order, p the number of oscillatory modes, and q the number of exponential modes. (a) Responses to step function disturbances with exponential modes only. Note that response may have an overshoot. (b) Exponential plus oscillatory mode with the latter having the smallest time constant. (c) and (d) contain two oscillatory modes having well-separated angular velocities and damping times.

3 DYNAMIC BEHAVIOR; THE TRANSIENT RESPONSE

a. Previous sections have described two essentially different modes of behavior exhibited by systems of linear processes when disturbed. The exponential mode of Fig. 3-4 is defined by its time constant, and the damped oscillatory mode of Fig. 3-9 by its angular velocity (ω) and damping time constant T. The term *mode of response* or *mode of free vibration* refers to either of these two types of behavior, by which a linear system readjusts itself to a disturbance or a change in the operating conditions. Note that although no oscillations appear in the exponential mode, the term — mode of free vibration — includes both forms.

b. These modes, which are characteristic of the system and not of the disturbance, are added to make up the transient portion of the complete response. In performing this summation, however, the amplitude and sign of the exponential modes are adjusted to satisfy conditions imposed by the location and nature of the disturbance, and by the initial state of the system (initial conditions). In a similar manner, oscillatory modes will have signs, amplitudes, and phase angles established by the disturbance and initial conditions.

c. Although the modes of free vibration are fixed for a given system, the manner in which the several modes may be combined is subject to wide variation, so that although the responses of two different dependent variables may contain exactly the same modes, to a superficial inspection they may not appear to have common components.

d. In counting the number of modes appearing in a given response, each oscillatory mode counts for two, since it is formed by two compartments or some equivalent process. With the total number of modes fixed at n, and with only two kinds of mode possible, the number of combinations is strictly limited, as noted in the caption.

e. With all modes exponential, the response may appear as in (a), and lie anywhere between the highly damped response without overshoot and the curve with one overshoot. With the existence of oscillatory modes a wide variety of responses is possible. Part (b) shows the response of a third-order system having one oscillatory and one exponential mode. The responses in (c) and (d) each contain two oscillatory modes, but of quite different damping time constants.

f. Although system order increases with the addition of compartments, no new kinds of modes appear. Regardless of the system order, the only possible modes in a linear system are the exponential and damped oscillatory modes previously described.

3.13 MODES OF FREE VIBRATION   71

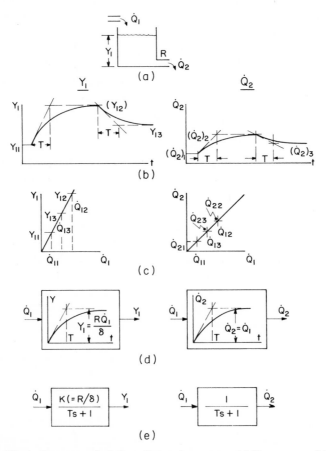

Fig. 3-13. Block diagram symbols for a first-order process. (a) The process. (b) Responses to step function inputs. (c) Steady-state characteristics with corresponding operating points. (d) Block diagrams with step function response shown therein. (e) Transfer functions.

3   DYNAMIC BEHAVIOR; THE TRANSIENT RESPONSE

a. Although the block diagrams shown so far have been restricted to the steady-state relations, they can readily be extended so as to provide both steady-state and dynamic information. The diagrams in this figure all relate to the single-compartment process whose dynamic behavior is given by the exponential function (Blesser, 1969).

b. Although this is a first-order process, the output variable could be selected as Y_1 or $\dot{Q}_2$. The temporal responses to changes in $\dot{Q}_1$ are shown in (b), and the successive operating points are indicated in (c). Note that the time constant for all the responses is the same, namely, T.

c. The block diagrams in (d) characterize the component by its response to a step function change in the input. Since the response is an exponential curve of time constant T, and the final steady-state values are given, the dynamic character of the component is completely specified. From this information, the response to any other input signal could, in principle, be computed.

d. A second way of defining the dynamic characteristics of the block is by means of a *transfer function,* as shown within the boxes in (e). These expressions are in the Laplace transform notation, and the quantity s is the Laplace variable. Although means are available for calculating the response of the output variable for any change in the input variable, we are here concerned only with the physical interpretation of the quantities appearing in the transfer function. The denominator, $Ts + 1$, states that the transient is a single exponential term having the time constant T. The numerator gives the steady-state relation between the input and output quantities, such that multiplying the input by the numerator yields the final steady-state value of the output.

e. The block diagram and transfer function are ways of defining a process, and are widely used when some processes are coupled along with others to form a larger system. However, it should be noted that the coupling between processes must be unilateral if the system is to be represented by a number of individual blocks. Furthermore, when connected into a system, the input to a specific block will seldom be a step function as is assumed here, and in the feedback cases treated subsequently, the time constants in the response are not those of the individual processes.

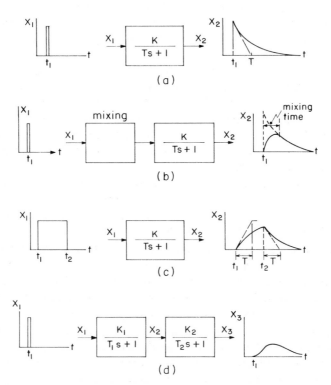

Fig. 3-14. Responses to impulse and pulse signals. (a) Impulse response of first-order process with instantaneous mixing. (b) Impulse response with mixing time. (c) Pulse response of first-order process. (d) Impulse response of a second-order system.

a. In most of the previous figures, a step function disturbance has been shown for which the response has both transient and steady-state components characteristic of the system and the disturbance. In a number of instances, it is more significant to treat the response to a pulse disturbance. This might be the case in pharmacological studies in which some substance is injected into the blood stream, or, possibly, ingested.

b. An *ideal impulse* is defined as a pulse of vanishingly small width and great height. Although it cannot be realized in practice, it can be well-approximated by a pulse whose width is small compared to the system time constants. The essential notion is that the material introduced as a disturbance shall be injected before the processes have time to change or react.

c. An "impulse" is the input signal to a first-order process in (a). If this process was a tank containing liquid, as in the previous examples, and the output variable was the outflow, then a disturbance consisting of adding a small amount of liquid to the tank (an impulse) would produce the response shown. The outflow would reflect the sudden change in material stored, and would have the decaying exponential characteristic of a first-order process.

d. The situation is different if the material added is a solute, or tracer, because in such circumstances a period is required for mixing. This accounts for the rising portion of the response curve in (b).

e. With a pulse of finite width, there will be transients associated with both the start of the pulse and its end. In the case shown, both portions would be exponential with the same time constant T.

f. With several compartments in cascade as in (d), the signal is propagated from one to the next, each compartment modifying the signal in its own way. The output of the first compartment (X_2) would be of the same form as the response in (a), and X_3 is the response of a first-order process when it receives an input of this exponential pulse character.

g. Theoretically, the pulse disturbance has certain advantages as a test stimulus. The response contains the sum of the exponential terms produced by each of the compartments, but there is considerable mathematical difficulty in separating them when the number is three or more. Furthermore, since each compartment attenuates the response, it becomes difficult to disturb the input with an impulse of great enough amplitude to provide a clearly observable signal at a point several compartments removed.

3.15   RESPONSES TO PULSE DISTURBANCES   75

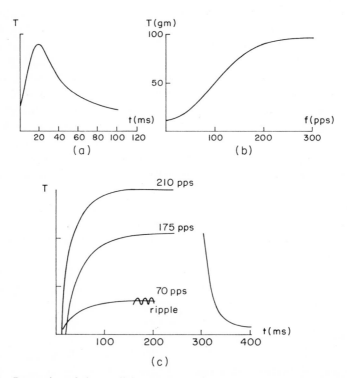

Fig. 3-15. Properties of the medial rectus muscle (of a cat) measured under isometric conditions. (a) Twitch response to a supramaximal stimulus. (b) Steady-state characteristic giving the isometric tension for a range of stimulus frequencies (Cooper and Eccles, 1930). (c) Step function response curves for synchronously firing motor units. The curves represent average tension, but there is an appreciable ripple component at the lower frequencies as indicated (Buller and Lewis, 1965).

a. The development of tension in a muscle following a single pulse stimulus, termed the *twitch response,* is shown in (a). The internal rectus is one of the fastest-responding muscles; others will be much slower to contract. The twitch response is obtained with a supramaximal stimulus delivered to the motor nerve or the muscle itself, in which all motor units participate, and contract simultaneously.

b. The character of this response, as well as all other muscle contractions, depends to a great extent upon a number of mechanical factors, including initial length, whether isometric or isotonic, and the parameters of the limb or mechanical load.

c. The steady-state behavior of a muscle is represented by the curve in (b). Over the intermediate range of frequencies, the tension developed is almost a linear function of frequency, but saturation occurs at the higher frequencies.

d. The development of tension following a step function change in stimulus frequency is shown in (c). In a sense this represents the summation of twitch responses, but the summation does not adhere strictly to the superposition principle. The ripple component, very obvious at low frequencies, is an artefact introduced by the synchronous firing of all motor units, and is not a factor under physiological conditions when the firing is asynchronous.

e. The step function responses in (c) may be closely approximated by exponential functions, and the initial rate of rise increases roughly in proportion to the stimulus magnitude, that is, the stimulus frequency (see Problem 3-7). Furthermore, even though the curves exhibit the effects of saturation as would be expected in view of (b), the initial rate of rise continues to rise with increasing stimulus frequency.

f. The fall in tension at the termination of the stimulus is frequently somewhat faster than the rise in tension at the beginning. This lack of symmetry between the rising and falling portions of the response is suggestive of nonlinearities within the system. Such evidence is not conclusive, but is enough to warrant further examination. See Section 9.14.

g. The above evidence points to the selection of a single time constant model for the muscle, possibly adequate for small signals, and essentially isometric conditions. Of more immediate interest is the fact that a muscle itself has a well-damped response with no sign of overshoot and oscillation, a behavior quite different from that sometimes found in a feedback control system as seen in Figs. 3-10 and 5-13.

a. The dynamical behavior described in this and subsequent chapters may be calculated mathematically by following the procedures outlined below. The steps listed are essentially the same for either a classical method of analysis (Guillemin, 1953) or for the Laplace transform method (Milsum, 1966).

b. A set of simultaneous differential equations is written from the underlying physical laws.

$$dx_1/dt = a_{11}x_1 + a_{12}x_2 + a_{13}x_3 + c_1 + e_1[t],$$

$$dx_2/dt = a_{21}x_1 + a_{22}x_2 + a_{23}x_3 + c_2 + e_2[t],$$

$$dx_3/dt = a_{31}x_1 + a_{32}x_2 + a_{33}x_3 + c_3 + e_3[t].$$

The above three equations represent three first-order components, with each one coupled to the other two via the coefficients a_{jk}. This would be the most general case. The terms c_k are constants, and the $e_k[t]$ are arbitrary functions of time; both of these terms might represent inputs to the compartments.

c. This set of n equations in n variables is reduced to a single differential equation of the nth order in (any) one of the dependent variables x_k. This may be done by means of Cramer's rule or other algorithm, yielding an equation of the form

$$b_n (d^n x_k/dt^n) + b_{n-1} (d^{n-1}x_k/dt^{n-1}) + \cdots + b_1 (dx_k/dt) + b_0 x_k$$

$$= C_k + E_k[t].$$

The terms on the right-hand side are the inputs or forcing function, and are independent of the variable x_k. This is the *equation of motion* for the variable x_k, whose solution will completely describe the behavior of this variable for given initial conditions, and in response to any disturbance to the system.

d. The complete solution when found will be of the form

$$x_k(t) = \text{steady-state solution} + \text{transient terms}$$

$$= (x_k)_{ss} \qquad\qquad + \sum_{k=1}^{n} (x_k)_t.$$

There will be one transient component for each order of the system. The transient terms describe the manner in which the system moves from its original state to its new steady state, and these terms disappear with time, leaving only $(x_k)_{ss}$, which is termed the *particular integral*.

e. The n transient terms are obtained by replacing the derivative terms in the equation of motion by the (complex) number s, to give the algebraic equation

$$b_n s^n + b_{n-1} s^{n-1} + \cdots + b_1 s + b_0 = 0.$$

Here s^k replaces the kth derivative, and the right-hand side is set equal to zero. This is the *characteristic equation*, which may be written in the form

$$s^n + (b_{n-1}/b_n)\, s^{n-1} + \cdots + (b_1/b_n)\, s + b_0/b_n = 0.$$

f. The n roots of this algebraic polynomial may be displayed by writing it in factored form:

$$(s - s_1)(s - s_2) \cdots (s - s_{n-1})(s - s_n) = 0,$$

where $s_1, s_2, \ldots, s_n$ are the roots (zeros). These must be found by appropriate algorithms if the detailed solution is sought. These n roots are either

(1) real numbers, s_k, or
(2) conjugate complex pairs, $s_i = \alpha_i + j\omega_i$ and $s_j = \alpha_i - j\omega_i$.

g. The transient terms in the solution are formed as follows. Each real root yields a term of the form $\exp(s_k t)$, and each pair of complex roots a term of the form $\exp(\alpha_i t)\,\sin(\omega_i t + \psi_i)$. The transient terms in the complete solution are then

$$\sum_{k=1}^{n} (x_k)_t = \sum_{k=1}^{m} B_k \epsilon^{s_k t} + \sum_{j=1}^{(n-m)/2} C_j \epsilon^{\alpha_j t} \sin(\omega_j t + \psi_j),$$

where B_k, C_j, and ψ_j are the constants of integration to be calculated so that the complete solution satisfies the initial conditions.

h. The complete solution is written as

$$x_k = (x_k)_{ss} + \sum_{k=1}^{m} B_k \epsilon^{s_k t} + \sum_{j=1}^{(n-m)/2} C_j \epsilon^{\alpha_j t} \sin(\omega_j t + \psi_j).$$

Two points should be noted. There are m constants B_k, and $(n-m)/2$ of each of the constants C_j and ψ_j; thus a total of n integration constants. These may be calculated in a number of ways. The second point is that for the normal stable system, all s_k and α_k are negative so that the exponential terms all decrease and disappear with time.

i. The n integration constants provide the means by which the general solution is "adjusted" to fit a specific physical problem. Inasmuch as we are not immediately concerned with the computation of complete solutions, further discuss of initial conditions is minimal. However, although the solution of a linear equation has the same general form for all sets of initial conditions, this is not true of nonlinear equations, for which the qualitative character of a solution may change radically with the initial conditions.

Epitome 3.18

1. The principle of superposition, applicable to linear systems, permits one to separate the dynamic response into steady-state and transient components. The latter is short-lived and represents the transition of the system in moving from one steady state to another.

2. A compartment or tank accumulates material if the inflow rate exceeds the outflow. The accumulation is the integral over time of the difference between inflow and outflow rates. Conversely, material is lost from the compartment when the outflow exceeds the inflow.

3. If the loss function is such that the outflow rate is proportional to the quantity of material stored within the compartment at each instant of time, the process is linear, and its transient behavior is described by an exponential curve.

4. The exponential transient has a slope that is initially large and decreases to zero as time increases, the slope at any instant being proportional to the difference between the final and the present value. The decaying exponential curve is completely characterized by its initial (or present) value and a time constant.

5. The response of a first-order process to a step function disturbance may be expressed as the algebraic sum of a constant (the steady state) and an exponential transient. The time constant describing the latter is the same for all dependent variables.

6. A step function disturbance to a first-order process results in an exponential transient having a time constant given by the conditions after the disturbance occurs. Although the dynamic behavior of the several dependent variables may appear to be quite different, they all contain the same components, but with different signs and amplitudes.

7. The dynamic response of a second-order system contains two exponential transient terms, as well as the steady-state. If the processes are noninteracting, the time constants describing the transient components are individually obtained from the two compartments.

8. Noninteracting components yield the same dynamic behavior ir-

respective of the sequence of the several processes. The time constants describing the behavior of interacting processes are not associated with the individual compartments, but must be calculated from an expression involving all the system parameters.

9. When signals are transmitted through a series of compartments, each compartment receives, as an input signal, the output signal from the previous compartment. As a consequence, the signal tends to be progressively attenuated and delayed in time as it progresses through the system.

10. Systems of the second order and higher may exhibit transient oscillations when disturbed. These are found in all types of physical systems if it is possible for energy to flow back and forth between two sites or forms (kinetic and potential).

11. Oscillations are not normally observed in the knee jerk, but may be pronounced in cases of cerebellar dysfunction (dysmetria). In general, neuromuscular control appears to be well damped, but oscillations are readily observed when the damping mechanisms (presumably in the cerebellum) are upset.

12. Systems of many components or processes yield high-order responses containing a number of transient components equal to the order of the system. System order is equal to the number of compartments coupled to each other in a given disturbance–response experiment.

13. High-order systems may exhibit both exponential and damped oscillatory transient terms. These modes of free vibration may appear in many combinations.

14. The block diagram symbol for a first-order process contains the steady-state gain and the time constant, which completely define the behavior of a single compartment. The response to any type of input may be calculated from this information.

15. The ideal impulse response of a first-order process is a single time constant, exponentially decaying, curve if no mixing time is required. In the presence of some mixing process, time is required for the output to build up to its maximum value.

16. The isometric response of a skeletal muscle to a step function change in stimulus frequency is approximately represented by a first-order process. However, the initial rate of rise of tension continues to increase with stimulus frequency even though the steady-state tension saturates. Furthermore, there is some evidence of asymmetry between the rise and fall in tension.

17. The mathematical analysis of a linear system involves the following steps:

    (a)   Write a set of differential equations describing the physical processes.

(b) Form a single differential equation of the *n*th order from the set of equations in (a).

(c) Find the roots of the characteristic equation.

(d) Find the steady-state solution of (b).

(e) Write the complete solution, including the steady-state and transient terms.

(f) Compute the constants of integration from the known physical conditions at $t = 0$.

The precise nature and sequence of these several steps will vary with the method used to obtain the solution.

Problems

1. Repeat the construction of the dynamic behavior shown in Fig. 3-4 for the case in which the process is initially in a steady state, $\dot{Q}_1 = \dot{Q}_0$, and R is then suddenly reduced to one-half its former value.

2. The first-order system of Fig. 3-5 is operating in the steady state, with $\dot{Q}_1 = \dot{Q}_2$, when a fixed quantity of liquid is suddenly added to that in the tank. The variable Y_1 then suddenly jumps by the amount ΔY_1. Sketch the ensuing behavior of Y_1 and $\dot{Q}_2$.

3. Two compartments are coupled in a noninteracting manner as in Fig. 3-6. The upper compartment has the time constant T_1 and the lower one a time constant T_2.

(a) Assume $T_1 > T_2$. The compartments are initially empty when $\dot{Q}_1$ is started and maintained at a constant rate. Sketch the behavior of Y_1, $\dot{Q}_2$, Y_2, and $\dot{Q}_3$ until the system reaches a steady state.

(b) With the system in the above steady state, R_2 is suddenly decreased a small amount $(-\Delta R_2)$. Sketch the ensuing behavior of the system variables.

4. The response of a second-order system to a step function disturbance is found to be represented by the expression

$$x[t] = 10 - 12 \ exp(-t/5) + 2 \ exp(-t/2).$$

(a) Plot $x[t]$. (Hint: draw each of the two exponential terms as a straight line on semilog paper and add corresponding ordinates.)

(b) How many seconds are required for the response to attain to within 1% of its final value? Which of the terms in the above expression largely determines this final approach to a steady state?

5. The transient component of an observed response is tabulated below.

t(sec)	Y_t(cm)	t(sec)	Y_t(cm)	t(sec)	Y_t(cm)	t(sec)	Y_t(cm)
0	8	6	5.9	12	4.4	18	3.25
2	7.3	8	5.4	14	4.0	20	2.96
4	6.6	10	4.85	16	3.6	∞	0

(a) Is this an exponential curve? What is the time constant?

(b) The single-compartment system to which this applies is initially empty. $\dot{Q}_i$ is suddenly raised to 50 cm³/min and kept at that value. If the final value of Y is 8 cm, what is the magnitude of δ/R? What is the magnitude of A?

(c) Sketch the complete response for (b).

6. The transient component Y_t as observed for a given system is given below.

t(min)	Y_t(cm)	t(min)	Y_t(cm)	t(min)	Y_t(cm)	t(min)	Y_t(cm)
0	4	4	2.45	10	0.82	18	0.17
1	3.65	5	2.1	12	0.56	20	0.11
2	3.27	6	1.7	14	0.37	∞	0
3	2.9	8	1.2	16	0.25		

Does this consist of one, or more, exponential curves? What are the time constants?

7. Consider the step function response of a first-order process.

(a) Show that the initial rate of change (fall or rise) is proportional to the magnitude of the input step.

(b) If the initial rate of change is to remain constant, how must the time constant be adjusted with the magnitude of the input step function?

References

Blesser, W. B. (1969). "A Systems Approach to Biomedicine." McGraw-Hill, New York.
Buller, A. J., and Lewis, D. M. (1965). Rate of tension development in isometric tetanic contractions of mammalian fast and slow skeletal muscles. *J. Physiol.* **176**, 337–354.

Cogan. D. G. (1954). Ocular dysmetria, flutter-like oscillations of the eyes and opsoclonus. *AMA Arch. Ophthalmol.* **51,** 318.

Cooper, S., and Eccles, J. C. (1930). The isometric responses of mammalian muscles. *J. Physiol.* **69,** 377–385.

Daniels, F. (1956). "Mathematical Preparation for Physical Chemistry." McGraw-Hill, New York.

Guillemin, E. A. (1953). "Introductory Circuit Analysis." Wiley, New York.

Holmes, G. (1922). On the clinical symptoms of cerebellar disease. *Lancet* **202,** 1177–1182.

Milsum, J. H. (1966). "Biological Control Systems Analysis." McGraw-Hill, New York.

Riggs, D. S. (1963). "Mathematical Approach to Physiological Problems." Williams & Wilkins, Baltimore, Maryland.

Ruch, T. C., and Patton, H. D. (1965). "Physiology and Biophysics," 19th ed. Saunders, Philadelphia, Pennsylvania.

Trimmer, J. D. (1950). "Response of Physical Systems." Wiley, New York.

3   DYNAMIC BEHAVIOR; THE TRANSIENT RESPONSE

A comparison of the behavior of two systems, which are identical except for the presence of a feedback mechanism in one and its absence in the other, is helpful for the development of an understanding of feedback processes. The consideration of changes brought about by the addition of feedback, which may be termed an engineering approach, has the merit of isolating those effects clearly attributable to feedback. However, from the viewpoint of biology, this may seem somewhat contrived, since the possibility of observing any one system functioning with and without feedback is somewhat remote. Although such a comparison can be made in a few cases, for the most part the opportunity does not occur naturally, and technical problems make it difficult to accomplish experimentally.

The acquisition of feedback processes reaches far back into evolutionary history, and questions of how and where feedback evolved are well beyond any discussion contemplated in this volume. The fact that feedback can be identified in the lower organisms and in basic biochemical reactions suggests that it appeared early in evolutionary history (Adolph, 1968).

From a biological viewpoint, appropriate questions might take the following forms. Can feedback be demonstrated in this living system, and what components and pathways partake in the feedback process? If feedback can be identified, of what significance might this have in experimental investigations and the interpretation of experimental observations? In what ways does feedback alter the steady-state and dynamic behavior? The effects of feedback may be so pervasive that questions of this character can become highly important in developing an adequate understanding of living systems.

As one examines the variety of biological processes, from biochemical reactions within a cell to the behavior of an entire organism, it is difficult to find processes that are not parts of some feedback system. The ubiquity of feedback systems testifies to the importance of this mechanism to life, and lends further impetus to a study of its properties and consequences.

This chapter is restricted to a discussion of one type of feedback and to the steady-state effects it produces. Unfortunately, many discussions of biological feedback have been based upon definitions that are so broad that they include all stable physical systems, and thus are useless in refining the description of homeostatic processes. We shall begin with an operational definition that may prove to be too restrictive to encompass all types of feedback, but it has the advantage of providing a sound basis for discussion of fundamental principles.

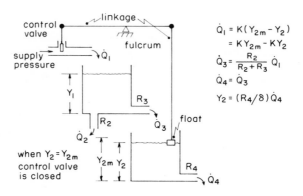

$$\dot{Q}_1 = K(Y_{2m} - Y_2)$$
$$= KY_{2m} - KY_2$$
$$\dot{Q}_3 = \frac{R_2}{R_2 + R_3} \dot{Q}_1$$
$$\dot{Q}_4 = \dot{Q}_3$$
$$Y_2 = (R_4/8)\dot{Q}_4$$

Consequences of a number of disturbances
in the absence of feedback

	Dependent variable			
Disturbance	$\dot{Q}_1$	Y_1	Y_2	$\dot{Q}_4$
$+\Delta$ supply pressure	$+\Delta$	$+\Delta$	$+\Delta$	$+\Delta$
$+\Delta R_3$	0	$+\Delta$	0	0
$+\Delta R_2$	0	$+\Delta$	$+\Delta$	$+\Delta$
$+\Delta R_4$	0	0	$+\Delta$	0

Fig. 4-1. Fluid flow system with feedback regulation of the liquid level Y_2. Any disturbance that affects the magnitude of Y_2 will cause $\dot{Q}_1$ to change in a direction to minimize the effect of that disturbance on Y_2. Because of feedback, a disturbance will generally change the value of all system variables. However, the table shows only the steady-state consequences of a number of disturbances *in the absence of feedback*.

a. Negative feedback, consisting of the float, linkage, and valve, has been added to a fluid flow system of the type previously discussed. This provides regulation for Y_2 such that if Y_2 decreases, the valve will open somewhat to increase $\dot{Q}_1$, and vice versa.

b. Presumably, Y_2 is critical for this process, and it is this variable that is used to control $\dot{Q}_1$; Y_2 is termed the *regulated variable*. All physical processes are continually beset by a host of disturbances and fluctuations that tend to displace the steady-state operating point, and thus the magnitude of Y_2. Regardless of the character and location of the disturbance, if it affects Y_2 the regulating process will act to minimize the changes in the regulated variable.

c. The figure shows the steady-state equations that relate the 4 variables. The first equation describes the feedback mechanism as a linear relation between Y_2 and $\dot{Q}_1$, in which $\dot{Q}_1$ is seen to decrease as Y_2 increases. Y_{2m} is a fictitious value of Y_2 at which the valve would be completely closed, and the constant K is the gain of the feedback mechanism.

d. Changes in the supply pressure, or in the quantities R_2, R_3, R_4, might well be expected, and any such disturbance will cause a change in Y_2. The table shows what the steady-state effects would be in the absence of feedback. With the feedback effective, however, all the variables are interrelated, and all will change somewhat for any of the disturbances. These changes cannot be found by simple inspection, but require a more detailed analysis as described subsequently. The addition of a feedback connection produces an interrelation between all the system variables so that not only are the several flows coupled to each other in the "forward" direction, but there is now a coupling between the regulated variable Y_2 and $\dot{Q}_1$.

e. The regulatory action described above was said to minimize the effect of disturbances, which leaves open the question of whether feedback can completely eliminate the effects of disturbances. Although there are some regulatory systems in which the regulated variable can be held at a fixed value, this is not the case here (see Section 4.15). The present system acts to minimize the consequence of disturbing factors, and by suitable design may reduce such effects to negligible proportions.

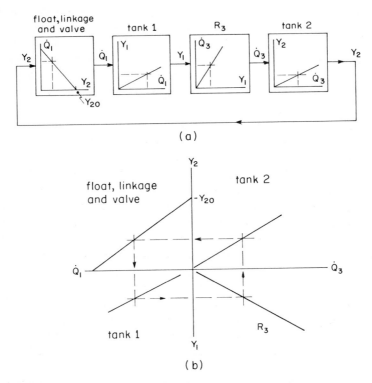

Fig. 4-2. Graphical construction for finding the steady-state solution, or operating point, of the system in Fig. 4-1. (a) Block diagram. (b) Graphical solution in which the steady-state characteristic of each of the four components is plotted in a separate quadrant. The dotted lines show the values of the several variables that simultaneously satisfy the four steady-state charactertistics.

a. This figure is drawn to show the steady-state relations for a negative feedback system similar to that in Fig. 4-1. Each of the four components is described by its own steady-state characteristic curve.

b. The "output" of each component serves as the "input" to the next in sequence, and the variables are said to be cyclically related. The dotted lines serve to depict the *operating point,* defined by the magnitudes of the four variables such that the above cyclic relations between them are satisfied. These values of the four variables are said to simultaneously satisfy the four steady-state characteristics (Adolph, 1943).

c. If the four steady-state characteristic curves are plotted as in (b), the operating point may be found by trial. In this system, and in all strictly linear systems, there is only one possible operating point. However, this is not true for many nonlinear systems, for which more than one operating point is possible.

d. Note that the steady-state magnitudes of the four variables depend upon all four characteristics, and a change in any one of them will result in a change in all four variables. It is this intimate relationship between one component in the system and the behavior of all the system variables that makes feedback systems difficult to study (Jones, 1951, 1969).

e. With this figure in mind, a negative feedback system is defined in the following operational sense.

    (1) A set of processes is identified each with its own "input" and "output" variable.

    (2) These processes are coupled in a cyclic manner so that the output of one component forms the input to the next.

    (3) Among the processes there is one (or an odd number) that exhibits *sign reversal*, that is, the characteristic has a negative slope.

    (4) For effective feedback, at least one component should exhibit unilateral coupling. It may be shown that this is necessary to produce low sensitivity to disturbances.

f. The above rules serve as a means of identifying negative feedback. Although these rules may be inadequate in some cases, they do serve to isolate a broad class of systems having the properties usually associated with negative feedback. Note that the rules are applied to the physical system, not to its mathematical representation.

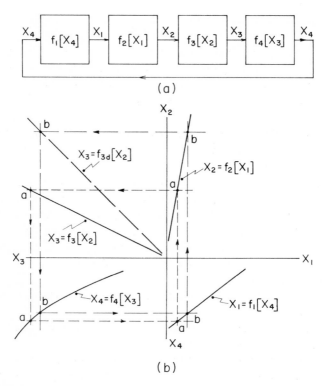

(a)

(b)

Fig. 4-3. Effect of a disturbance on the operating point of a feedback system. (a) Block diagram. (b) Graphical solution modified to show the effects of a disturbance which changes $f_3[X_2]$ to $f_{3d}[X_2]$.

4   INTRODUCTION TO FEEDBACK; THE STEADY STATE

a. The discussion of the previous section is continued in the present figure, so as to show the effects of a disturbance in more detail and to introduce several related concepts. This figure again shows a system of four components, but the variables have been changed to make it more general (Jones, 1960).

b. The functional notation $X_2 = f_2[X_1]$ is a terse way of denoting that X_2 is a function of X_1, and for every value of X_1 there is a corresponding value of X_2. The nature of $f_2[X_1]$ must be given in either graphical form (as in the figure) or in algebraic form before any calculations can be carried out. The square brackets will be used exclusively to denote such a functional relationship, which may be simply a steady-state characteristic as in this instance, or to denote a function of time in dynamic problems.

c. The solid characteristic curves in the four quadrants represent the undisturbed state, and the four points marked a designate the values of the four variables. The disturbance consists of a change from $f_3[X_2]$ to $f_{3d}[X_2]$, and the disturbed state is given by the points marked b. Note that all variables have been perturbed — some more than others — and that some variables have increased whereas others have decreased, all as a consequence of a single disturbance. The disturbance shown here amounts to a change in gain, but other more general types of change in the functions are possible.

d. Inasmuch as the operating point depends upon the shape and location of all the characteristic curves, so the perturbations produced will be functions of all the components. Another way of stating this is to note that although a change in f_3 produces a change in X_3, the latter perturbation affects all the succeeding variables and thus in turn X_2, which further modifies X_3. The perturbed operating point must simultaneously satisfy all the perturbed characteristics.

e. It should now be clear that the operating point is a multidimensional point, in that it is completely defined only by stating the steady-state values of all the system variables. It may be looked upon as a point in multidimensional space, whose coordinates are the values of the several variables.

f. The construction shown in this and previous figures has been confined to systems of only four components and four dependent variables. For systems of higher order, some of the characteristic curves may be combined, as in Fig. 4-6 and subsequent figures.

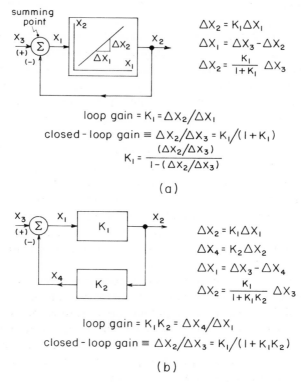

$$\Delta X_2 = K_1 \Delta X_1$$

$$\Delta X_1 = \Delta X_3 - \Delta X_2$$

$$\Delta X_2 = \frac{K_1}{1+K_1} \Delta X_3$$

loop gain $= K_1 = \Delta X_2/\Delta X_1$

closed-loop gain $\equiv \Delta X_2/\Delta X_3 = K_1/(1+K_1)$

$$K_1 = \frac{(\Delta X_2/\Delta X_3)}{1-(\Delta X_2/\Delta X_3)}$$

(a)

$$\Delta X_2 = K_1 \Delta X_1$$

$$\Delta X_4 = K_2 \Delta X_2$$

$$\Delta X_1 = \Delta X_3 - \Delta X_4$$

$$\Delta X_2 = \frac{K_1}{1+K_1 K_2} \Delta X_3$$

loop gain $= K_1 K_2 = \Delta X_4/\Delta X_1$

closed-loop gain $\equiv \Delta X_2/\Delta X_3 = K_1/(1+K_1 K_2)$

(b)

Fig. 4-4. Open-loop and closed-loop gains of a feedback system. (a) System having unity feedback for which $X_1 = X_3 - X_2$. The loop gain K_1 may be calculated from the closed-loop measurement of $\Delta X_2/\Delta X_3$. (b) System having nonunity feedback for which $X_1 = X_3 - K_2 X_2$. The loop gain $K_1 K_2$ cannot be calculated from the closed-loop gain $\Delta X_2/\Delta X_3$ alone.

a. Many, but not all, feedback systems may be represented by the block diagram in (a), where the entire regulated process is represented by the single block in the forward path and the regulated variable X_2 is subtracted from the *reference input* X_3. The latter is a constant quantity that serves to establish the steady-state value of X_2. Note that the quantities X_2 and X_3 must be alike physically.

b. The system in (a) is said to have *unity feedback*. The *closed-loop gain* is simply $\Delta X_2/\Delta X_3$, and is given in terms of the *loop gain* K_1 by the expression shown. Note that K_1 must be dimensionless, that is, X_1, X_2, and X_3 must all be the same kind of physical quantity.

c. The system in (b) has a component in the feedback pathway, and is said to have *nonunity feedback*. Thus, although K_1 and K_2 need not be dimensionless, their product must be, so that variables X_4 and X_3 will have the same dimensions and can be subtracted at the *summing point*. The component in the feedback path is frequently a transducer, one of whose functions is that of transforming the regulated variable to one having the dimensions necessary for subtraction at a summing point.

d. As already indicated, the properties of a feedback system are intimately related to the magnitude of the loop gain, and thus it is frequently necessary to know or to measure its magnitude. In theory, the gain could always be measured by opening the loop, but in many instances it is impractical to do so. The question then arises: can the loop gain be obtained from closed-loop measurements? In the case of the unity feedback system in (a), this is possible, as shown in the figure. The situation is quite different for the nonunity feedback case in (b), where it will be seen that the expression for $\Delta X_2/\Delta X_3$ cannot be solved for the loop gain K_1K_2.

e. Examination of the expressions for the closed loop, that is, $\Delta X_2/\Delta X_3$, will show that this ratio approaches unity as K_1 is made large. This is frequently the case, especially in technological regulators, and it is in this sense that the reference input X_3 may be said to establish or "set" the operating point.

f. The summing point in the diagrams represents a subtraction process which physically may be performed in a variety of ways. Many of the components that perform the functions of the summing point and reference input do not do so in as explicit a manner as may be suggested by this figure. A specific physical case is treated in the next section.

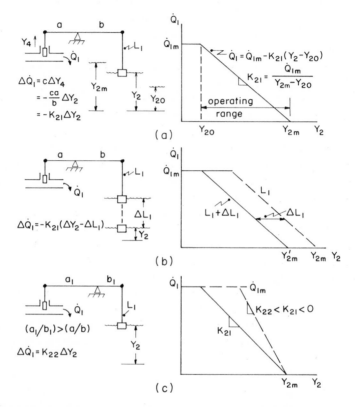

Fig. 4-5. Analysis of the control mechanism from Fig. 4-1 showing the factors affecting the operating point and loop gain. (a) The operating range is limited by the maximum flow rate ($\dot{Q}_{1m}$) and the level required to close the valve (Y_{2m}). The flow rate $\dot{Q}_1$ is a linear function of valve position Y_4; the constant of proportionality is c. (b) An increase in length of link L_1 reduces the steady-state value of Y_2 for a given Q_1. (c) An increase in the lever arm ratio (a/b) increases the loop gain.

4   INTRODUCTION TO FEEDBACK; THE STEADY STATE

a. The control mechanism shown in Fig. 4-1 serves to increase the magnitude of $\dot{Q}_1$ as Y_2 falls, and thus aids in regulating Y_2. However, the precise manner in which the regulatory functions depend upon the parameters of this mechanism may not be evident from a cursory examination of the figure.

b. Physical aspects which serve to set limits for the operating range of this control mechanism are shown in (a). An upper limit to the magnitude of Y_2 is set by that value which closes the valve; this is Y_{2m}. On the other hand, when Y_2 drops to Y_{20} the valve is completely open, $\dot{Q}_1$ is at its maximum value $\dot{Q}_{1m}$, and no further control of the inflow is possible.

c. Between these two extremes Q_1 is a linear function of Y_2, but beyond these limits regulation of Y_2 is impossible. This is another example in which physical limitations of the equipment establish operating limits, between which satisfactory operation must be confined. One should expect to find similar limitations imposed by physiological components. The expression for $\dot{Q}_1$ in (a) is readily derived (Problem 4-9).

d. The expression for $\dot{Q}_1$ contains a constant, written either as $\dot{Q}_{1m}$ or $K_{21}(Y_{2m} - Y_{20})$, a value derived from the dimensions of the equipment used. This quantity may be termed a *generalized reference input;* the adjective "generalized" is added to denote that this quantity is implicit in the control mechanism, and does not represent an existing physical quantity (flow rate) from which the output is substracted.

e. We next examine the effects of various changes in the control mechanism. In (b), the length of the link L_1 is increased. This has the effect of changing the level Y_2 associated with a given $\dot{Q}_1$ and thus of changing the steady-state value of Y_2. In terms of the previous expressions, an increase in L_1 decreases Y_{2m}.

f. The ratio of the lever arms a/b can be changed by moving the location of the fulcrum. The consequence is to change the slope of the characteristic, and thus the gain associated with the control mechanism. This modification does not alter Y_{2m}, but does change the reference input $K_{21}(Y_{2m} - Y_{20})$ accordingly.

g. This example should have shown that the functions of reference input, sign reversal, and gain may be interrelated so that the change in a single physical quantity may have more than one effect upon the system. We conclude that the conventional block diagram of Fig. 4-4 may well be misleading when applied to an actual system, especially when the components are physiological.

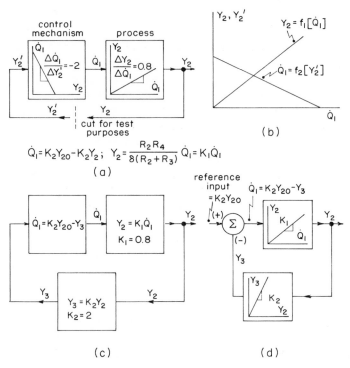

Fig. 4-6. Definition of loop gain. (a) Block diagram showing system cut at Y_2, Y_2'. Loop gain is $\Delta Y_2/\Delta Y_2' = (\Delta \dot{Q}_1/\Delta Y_2')(\Delta Y_2/\Delta \dot{Q}_1) = (2)(0.8) = 1.6 = K$. (b) Graphical construction for finding the operating point. (c) Block diagram with gain and sign reversal functions of the control mechanism separated. (d) Block diagram with conventional summing point.

a. The concepts of open- and closed-loop gain, together with that of the operating point, are brought together here with particular reference to the regulating system of Fig. 4-1. The block diagram in (a) represents the system shown in Fig. 4-1, but reduced so as to show only two components, one having a characteristic with negative slope, and the other positive. The constants are designated somewhat differently than in Fig. 4-1.

b. A "cut" has been introduced at Y_2, Y_2' so as to permit measurement of the loop gain, that is, of $K = \Delta Y_2/\Delta Y_2'$. K is also equal to the product of all the individual component gains. This system assumes its normal condition when the two terminals Y_2 and Y_2' are joined.

c. Any negative feedback system can be resolved into just two blocks as in this figure. The graphical construction in (b) shows the two characteristics depicted in a more general form, that is, as $f_1[\dot{Q}_1]$ and $f_2[Y_2']$, with the operating point given by the intersection. Note that the line for $\dot{Q}_1 = f_2[Y_2']$ has been inverted to suit the axes shown.

d. The loop gain K as given in most engineering texts is written without the negative sign, that is, it is the absolute value of the ratio, denoted $|\Delta Y_2/\Delta Y_2'|$. The gain as given here (with the negative sign) is termed the *homeostatic index* by Riggs (1963, p. 100). However, we shall retain the engineering usage, so that the homeostatic index becomes $-K$.

e. Part (c) of the figure shows the same system that appears in (a), but with the gain and sign reversal functions of the control mechanism separated. The gain associated with this mechanism appears in the feedback path with its output denoted by Y_3. The quantity K_2Y_{20} is a constant associated with the control mechanism, and has the physical significance of being equal to Y_3 when the valve is completely closed. Typical values for the several gains have been shown for illustrative purposes.

f. The conventional block diagram is shown in (d), in which the sign reversal function is denoted by the summing point, and K_2Y_{20} serves as a "reference input." The utility of a reference input in technological regulators lies in the fact that components can be chosen for this function that have a long time magnitude constancy that exceeds that of the other physical components.

g. In biological systems, the function of reference input may be difficult to identify, and in most instances is probably not associated with the sign reversal. When this latter function is performed by a component having a characteristic with negative slope, it is said to constitute an *implicit summing point*.

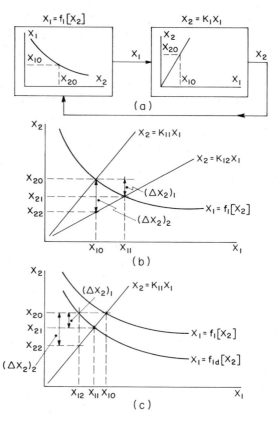

Fig. 4-7. Reduction in the effect of a given disturbance by virtue of feedback. The consequences of a disturbance are shown for the system both with and without feedback. (a) Block diagram. (b) Effect of a reduction in loop gain from K_{11} to K_{12}. (c) Effect of a disturbance to the sign reversal component.

a. A prime reason for adding feedback to a physical system is that by so doing, the effect of a variety of disturbing factors can be minimized and reduced to almost negligible values. This is an important engineering factor, and there is every reason to suppose that it is also significant biologically. Feedback serves to make an organism relatively immune to environmental changes.

b. The block diagram in (a) is essentially the same as those previously discussed. The lefthand component is nonlinear and includes an implicit summing point as indicated by its negative slope. The other block might represent any number of components in series, the overall characteristic having a positive slope. The operating point is given by the intersection of the two curves as before.

c. A disturbance which can affect the steady state of this system will appear as a change in one or the other of these two characteristics. For instance, assume that the gain of the second component changes from K_{11} to K_{12}. The new steady state is given by the coordinates X_{11}, X_{21}, and the change in X_2 is $(\Delta X_2)_1$.

d. Has feedback reduced the effect of this disturbance? This question may be answered by considering the situation in the absence of feedback. For the same value of X_2, that is, $X_2 = X_{20}$, the input to the second component was X_{10}. With this input fixed, and in the absence of feedback, the disturbance would reduce X_{20} to X_{22}, and the change in X_2 would be $(\Delta X_2)_2$. Feedback has reduced the effect of this disturbance to about half.

e. A similar situation exists if the disturbance affects the first component and not the second. Thus, if a disturbance causes $f_1[X_2]$ to become $f_{1d}[X_2]$, the operating point will move to X_{21}, X_{11}, but in the absence of it is X_{22}. As in the previous case, negative feedback has reduced the consequences of a disturbance compared to what it would have been in the absence of feedback.

f. In the above examples, the result of feedback has been a reduction in the effect of a disturbance by a factor of only two or so. This is a very modest reduction in comparison with values of 10 to 100 obtained in technological regulators. The extent of the reduction is determined by the shape of the curves, and in particular with the loop gain. These characteristics have a very low gain and hence will not demonstrate a large reduction, but they are adequate to depict the qualitative effects of feedback.

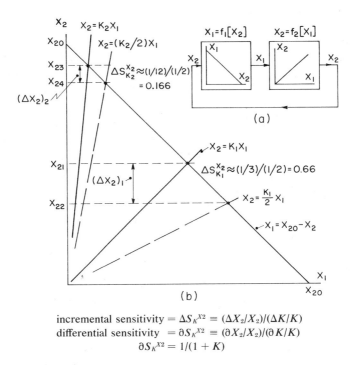

$$\text{incremental sensitivity} = \Delta S_K{}^{X2} \equiv (\Delta X_2/X_2)/(\Delta K/K)$$
$$\text{differential sensitivity} = \partial S_K{}^{X2} \equiv (\partial X_2/X_2)/(\partial K/K)$$
$$\partial S_K{}^{X2} = 1/(1+K)$$

Fig. 4-8. Incremental and differential sensitivity defined for a feedback system. (a) Block diagram. (b) Graphical construction illustrating the calculation of incremental sensitivity for two values of loop gain. In each instance the loop gain is assumed to be disturbed so as to reduce its value by a factor of one half.

a. Disturbances which arise in the environment or within the organism may affect the system in a variety of ways. The effects so produced are termed *perturbations,* and this section is concerned with the role of loop gain in reducing them.

b. In general, the larger the loop gain, the less effect a given disturbance will have on the regulated quantity. The disturbance considered here is a change in the magnitude of the gain. When $K = K_1$, the operating point is at X_{21}. A disturbance which reduces K_1 to $K_1/2$ causes the operating point to move to X_{22}. The perturbation is thus $(\Delta X_2)_1$.

c. On the other hand, if K were initially much greater ($K = K_2$), the operating point would become X_{23}, and a similar reduction in gain to $K_2/2$ results in the perturbation $(\Delta X_2)_2$. Thus, for similar disturbances the perturbation is greatly diminished if the loop gain is relatively high. Although there are important exceptions, the magnitude of a perturbation may be reduced by an increase in loop gain.

d. To describe the effect of disturbances in a quantitative manner, the following terms are introduced. ΔX_2 represents a small but finite change in X_2, and the ratio $\Delta X_2/X_2$ is the *per unit change.* The *incremental sensitivity,* $\Delta S_K^{X_2}$ is defined in the figure as the ratio of per unit changes in X_2 and K. The sensitivity for any other pair of quantities may be defined in an identical manner. Approximate values for the incremental sensitivity when $K = K_1 = 1$, and $K = K_2 = 10$ are shown in the figure.

f. In engineering literature, the sensitivity S is usually given as a differential sensitivity, here denoted as ∂S, and defined in the figure. The distinction between these two different sensitivities resides in the fact that the incremental value applies to changes of any size, whereas the differential sensitivity strictly applies only to infinitesimally small changes. A further comparison of these two definitions will be found in the next section (Horowitz, 1963).

g. For the single-loop linear system, ∂S is readily found to be $1/(1 + K)$.

h. The above definitions of sensitivity have essentially the same meaning as they do in normal discourse. An organism is sensitive to a given stimulus if it exhibits a pronounced response, whether that response is a perception or an overt action. With reference to homeostatic systems, the fact that the regulated quantity exhibits a low sensitivity does not imply that all variables behave the same. As a matter of fact, other variables must exhibit a large sensitivity if the regulated quantity is to remain nearly constant.

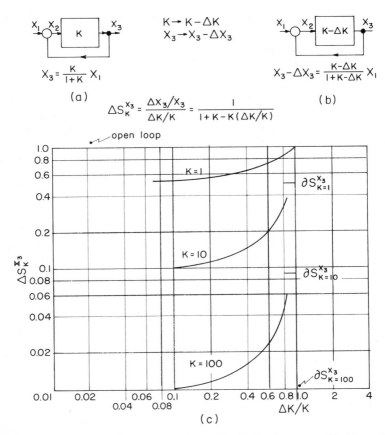

Fig. 4-9. Incremental sensitivity function for a unity feedback system with changes in loop gain. (a) Undisturbed system. (b) Disturbed system showing a reduction in loop gain. (c) Incremental sensitivity as a function of loop gain, and per unit change in gain. The corresponding differential sensitivity $\partial S_K{}^{X_3}$ is indicated at the right.

a. Many homeostats appear to be very effective regulators in that their sensitivity is quite low. Although it is not clear that biological systems achieve this low sensitivity by means of high loop gain alone, the matter is sufficiently important to warrant further discussion.

b. The previous section defined both differential and incremental sensitivity, but since disturbances are more likely to be relatively large, incremental sensitivity is believed to be the more useful concept. This quantity is plotted in the figure.

c. The disturbance considered is a finite reduction in K, that is, $K \rightarrow K - \Delta K$, and the corresponding change in X_3 has been calculated from the expressions shown for three different initial values of K. Without feedback, ΔS is always unity, but is reduced to small values with feedback if the loop gain is made high.

d. For very small disturbances, that is, $\Delta K/K \rightarrow \partial K/K \rightarrow 0$, the magnitude of ΔS approaches that of ∂S. This figure makes it clear that the incremental sensitivity is much larger than the differential sensitivity for large disturbances.

e. Technological regulators frequently have gains of 100 or more, and maintain sensitivities at values of 0.01 or less. However, at the present time there is insufficient information regarding homeostats to make a meaningful comparison. There can be little doubt that the sensitivity of homeostats is low, but there is considerable question that this is attained by high gain.

f. With few exceptions, satisfactory loop-gain measurements have not been reported for physiological regulating systems. This is largely due to technical problems associated with "opening the loop" without disturbing the properties of the individual components. In the absence of reliable values for loop gain, it is impossible to attribute low sensitivity to gain alone.

g. Another explanation of the observed low sensitivity may reside in the large number of parallel pathways found in many homeostatic systems. For example, the multitude of nerve fibers and muscle fibers making up one neuromuscular pathway may, in effect, employ the statistical properties of the population to reduce the total effect of individual variations.

h. Although the system described in this figure had unity transmission in the feedback path, the conditions are quite different when there are components in this path, as in Fig. 4-4b. It is then found that the sensitivity for changes in a component located in the feedback path is much greater than it is for disturbances in the forward path.

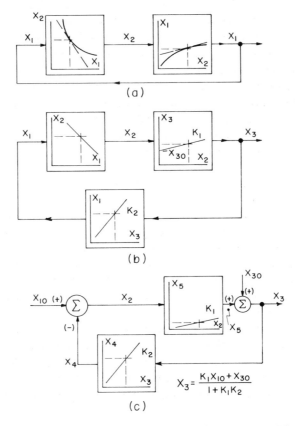

Fig. 4-10. Linear approximations to a nonlinear feedback system. (a) The nonlinear system, with (b) a linear approximation to the steady-state characteristics valid in the vicinity of the operating point. (c) Linear approximation with the conventional reference input and summing point.

a. The feedback regulator is generalized somewhat in this figure by showing both characteristics as nonlinear, with the operating point obtained in the usual manner as the simultaneous solution of the two curves, as in (a). If a tangent is then drawn through the operating points on the two curves, these straight lines form linear approximations to the actual curves.

b. Part (a) of the figure may be said to include an implicit summing point as described in Section 4.7. This characteristic has been resolved into two components in (b), in the same manner as shown in Fig. 4-6.

c. For some purposes it is sufficient to consider only small disturbances and therefore only small perturbations to the dependent variables. By thus restricting the perturbations, the relationships become linear and it is possible to compare this system behavior with that obtained from a purely linear one.

d. The block diagram shown in (b) still contains an implicit summing point, and this is replaced by a conventional one in (c). This part of the figure also differs from the one above by having an additional input X_{30} added to the variable X_5. This is another way of representing the intercept X_{30} in part (b). The expression for X_3 in (c) shows that the regulated variable now depends upon both X_{10} and X_{30}, so that the reference input X_1 has lost its previously unique position of largely setting the operating point for X_3.

e. It is now possible to make some comparisons between the strictly linear system of Fig. 4-4 and the more general nonlinear system in Fig. 4-10. Figure 4-4 shows that in the linear case, $X_2 = \{K_1/(1 + K_1)\}X_3$, so that as K_1 is made large the magnitude of X_2 approaches that of X_3. For instance, if $K_1 = 100$, $X_2 = 0.99X_3$, and in the steady state the magnitudes of X_2 and X_3 are nearly equal. In a real sense, X_3 establishes the operating point for X_2.

f. In the nonlinear case, two things are different. In the linear approximation in (c), there are two inputs, X_{10} and X_{30}, and the magnitude of the regulated variable depends upon both. Although X_{30} is in a sense a fictitious quantity whose value changes with the operating point, it is clear that X_{10} is no longer the sole determinant of X_2. Second, the linear approximation introduces nonunity feedback, and even if the loop gain were very large, X_3 in general would not become approximately equal to X_{10}.

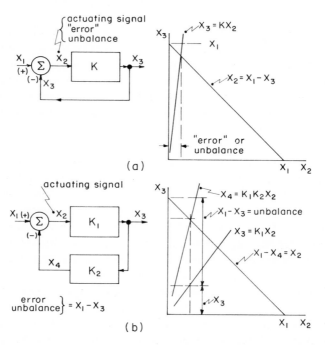

Fig. 4-11. The actuating signal in a feedback system is defined as the output of the summing point. (a) In a system with unity feedback the actuating signal can be identified with the error, or unbalance, since $X_2 = X_1 - X_3$. (b) With nonunity feedback the actuating signal is no longer the error.

4   INTRODUCTION TO FEEDBACK; THE STEADY STATE

a. Frequent reference to the error in a feedback system occurs in the engineering literature along with the postulate that it is the error that drives the regulated variable towards its desired value. This concept is of significance in servomechanisms, the principal function of which is to track or follow a given variable with the greatest possible precision, and the error becomes a measure of how well it succeeds. However, this concept may be misleading when applied to regulators or homeostatic systems.

b. The conventional unity feedback linear homeostat is shown in (a). If one defines the error as X_2, it is apparent that with high gain this quantity may be made as small as desired, as mentioned in the previous section. The error (or unbalance as it is sometimes termed) is amplified by the high gain system to produce the regulated quantity X_3.

c. The situation changes greatly with nonunity feedback as in (b), where with the same total gain as in (a), that is, $K_1 K_2 = K$, the actuating signal, $X_1 - X_4 = X_2$, can hardly be considered an error. Insofar as the summing point is concerned, X_4 is treated as if it were X_3, but minimizing the difference $X_1 - X_4$ is not the same as minimizing $X_1 - X_3$.

d. Similar conclusions would be reached for the nonlinear system in Fig. 4-10; the concept of identifying the error or unbalance as X_2 is quite misleading. As a matter of fact, some writers have introduced the normal value of X_3 into the block diagram as a fictitious input. By so doing, it is possible to generate a system error even though the diagram ceases to be isomorphic with the prototype. There is no evidence that a homeostat must contain within itself a physical quantity or information explicitly equivalent to the reference input of technological regulators. Note that this statement is inferred from limited evidence and some later investigation may reveal that an explicit reference input does indeed exist in a homeostat. However, it is believed that the reference is not necessary for the satisfactory functioning of a feedback regulator as long as the functions of sign reversal and adequate gain are present.

e. Even though an explicit error term may not be available in a homeostatic system, it is frequently desirable to describe the departure of the regulated variable from some appropriate norm. For this purpose a temporal or ensemble norm might be used as described later in Section 4.18 (Yamamoto, 1965).

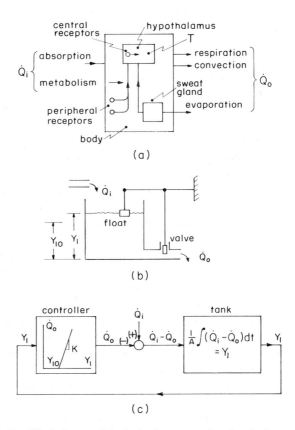

Fig. 4-12. (a) Simplified picture of the body thermostat showing the heat sources, and the pathways for heat loss. It is assumed that the body is under a heat stress so that loss by evaporation (sweating) is required to maintain a heat balance. (b) Liquid level regulator serving as an approximate analogue. (c) Block diagram of the liquid level regulator.

a. Overheating of the body from whatever cause will bring about sweating, and evaporation of this moisture is an effective means of regulating body temperature. Vasodilitation is another means of increasing heat dissipation, but it is assumed to be at a maximum effectiveness when sweating begins.

b. A simplified picture of the pathways involved in sweating is shown in (a). The body temperature remains constant only if the total heat loss (respiration, convection, and evaporation) is equal to the total heat influx (metabolism and heat absorbed from the environment). A large number of warm receptors located both centrally (hypothalamus) and peripherally (skin) provide the hypothalamus with signals indicative of body temperature, and appropriate signals are then sent to the sweat glands via neural (and possibly hormonal) pathways. The sensitivity of the system appears to be low; this indicates a large change in sweat rate for a small change in body temperature.

c. A much simplified and approximate hydraulic analogue appears in (b) in which $\dot{Q}_i$ and $\dot{Q}_0$ correspond to the heat flow rates, and Y_1 corresponds to the body temperature. In the steady state $\dot{Q}_i = \dot{Q}_0$, Y_1 is a constant, and $dY_1/dt = 0$. On the other hand, if $\dot{Q}_i \neq \dot{Q}_0$, then Y_1 is either increasing or decreasing.

d. The outflow $\dot{Q}_0$ will be zero if $Y_1 = Y_{10}$, where Y_{10} is simply that value of Y_1 at which the valve is closed. In the body thermostat, this would correspond to the threshold temperature at which sweating starts. For a given value of $\dot{Q}_i$, $Y_1 > Y_{10}$ by an amount just sufficient to make $\dot{Q}_0 = \dot{Q}_i$.

e. In this liquid level regulator, sign inversion occurs in the control of outflow by changes in Y_1. This is in contrast to the previous systems in which it was the inflow that was the manipulated quantity. Sign inversion is thus an inherent property of this system, although it may be shown explicitly in a block diagram.

f. The block diagram in (c) shows the relations between the several variables in more detail. The tank serves to integrate the *net* input $\dot{Q}_i - \dot{Q}_0$, and the value of the integral continues to change so long as the net input differs from zero. This would be true even though there were no feedback control mechanism; the latter simply serving to reduce the changes in Y_1 for a given disturbance.

g. The linear controller characteristic shown in (c) is distinctly a special case, and as a matter of fact with the float, level, valve arrangement shown is obtainable only by appropriate design of these several components. However, linearity is by no means a requirement for successful operation; see Fig. 9-9.

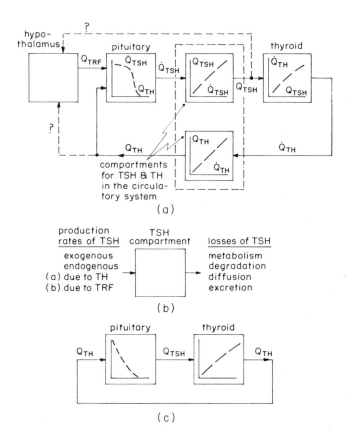

Fig. 4-13. Regulation of thyroid hormone (TH) concentration in the blood. (a) Each gland has a production rate $\dot{Q}$ that is a function of the concentration Q of the hormone produced by the other gland. The two dotted feedback paths to the hypothalamus are at present questionable. (b) The compartment for TSH in the circulatory system has the inputs and outputs shown and they jointly determine the concentration Q_{TSH}. A similar block diagram would represent the TH compartment. (c) Simplified block diagram merges each gland with the associated circulatory compartment.

a. Thyroxine (TH) participates in the regulation of many quantities in the body (metabolism, temperature), and the thyroid gland is thus a component in many feedback systems. In addition, there is feedback regulation of thyroxine concentration (Q_{TH}) in the blood brought about by the pituitary gland as suggested in (a) (Brown-Grant, 1969). The thyroid–pituitary system might be considered a minor feedback path in these other homeostats (Langley, 1965).

b. Although the output from each gland is a secretory rate, it is not directly measurable, so that it is convenient to use the concentration levels Q_{TH} and Q_{TSH} in the blood as the dependent variables (Distefano and Stear, 1969). The TH and TSH compartments are shown explicitly in (a), but are merged with the glands in (c). The TSH compartment, for example, has a number of inputs and outputs so that in the steady state, as shown here, Q_{TSH} is a function of Q_{TH}.

c. In the steady state, the total production and loss rates must be equal and they establish the concentration level in the blood. The exogenous production rate refers to the possibility of supplying the subject with additional TSH, whereas the endogenous rates are those associated with Q_{TH} and Q_{TRF}. TRF is a thyrotropin (TSH) releasing factor by means of which the hypothalamus presumably exerts control over the TH level.

d. Endogenous control of the rate $\dot{Q}_{TSH}$ consists of an inhibition of the pituitary secretion rate brought about by the circulating Q_{TH}. It is this inhibitory action that gives the pituitary gland a negative slope somewhat as shown (Purves, 1964). The operating point, at the intersection of the two characteristic curves, presumably lies in the region of the pituitary characteristic having a maximum (negative) slope.

e. The simplified diagram in (c) shows only two blocks, each one containing not only the gland itself but also the associated blood compartment. It is clear that this system meets all the requirements of a negative feedback regulator.

f. The simplified block diagram does not show the hypothalamus and will yield only one operating point. However, release of TRF by the hypothalamus, together with possible neural signals from the central nervous system, are believed to displace the pituitary curve and result in a new operating point. Furthermore, changes in thyroid function can conceivably move its characteristic, and also serve to change the operating point.

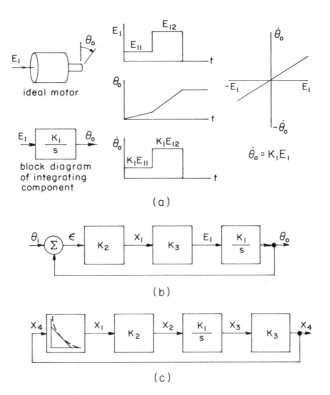

Fig. 4-14. Integrating component included in a feedback system to form a Type 1 regulator (servomechanism). (a) Ideal motor whose angular velocity $\dot{\theta}_0$ is proportional to the impressed voltage E_1 forms an integrating component, with the angular position of the output shaft as the output variable. (b) Type 1 system with the integrator output being the regulated variable. (c) A more general Type 1 system.

4   INTRODUCTION TO FEEDBACK; THE STEADY STATE

a. The previous examples have been limited to regulators in which disturbances bring about corrective changes but a deviation must remain in the regulated variable to maintain a continuing correction in the system. As a class, these are termed *proportional regulators,* although this term does not imply that the relations must be linear. We now turn to another class of regulating system termed the *servomechanism,* or *follower,* the function of which is to cause one physical variable to follow or track another.

b. The servomechanism contains a component described as a motor or integrating element having the characteristics shown in (a). This component has the property of producing a rate of change of the output variable that is proportional to the input. An ideal motor has this property: its angular velocity $\dot{\theta}_0$ is proportional to the input voltage E_1, and the output shaft continues to rotate as long as a voltage is impressed. This relation holds for both positive and negative values of E_1. The output shaft comes to rest at a particular value of θ_0 only when E_1 goes to zero.

c. The behavior of the complete system containing such an integrating element, together with the necessary amplifiers, may be inferred from (b). As long as ϵ and therefore E_1 are different from zero, the motor continues to rotate. The motor will stop (integration will cease) only when $\epsilon = 0$ and $E_1 = 0$. Thus, regardless of the magnitude of the input θ_i, the servo will continue to correct the output θ_0 until a zero error (actuating signal) results, and there is perfect correspondence between θ_i and θ_0. During the corrective action, the error may be positive or negative, and the servo motor must thus operate in either direction.

d. A servomechanism of the character described above is sometimes termed a *Type* 1 *system* to signify that it contains one integrating element within the loop designated as K_1/s. The proportional regulator is correspondingly termed a *Type* 0 *system.*

e. The integrating component that forms the heart of a servomechanism has the further property of being able to maintain any value of the output for zero input to the integrator. A device having this property is termed *astatic.* Although the servomechanism is applied to a wide variety of technological problems, it is extremely doubtful if biological counterparts exist.

f. In (c), the conventional summing point is replaced by a component having a negative slope, but a steady state can only be reached if $X_2 = 0$. This imposes restrictions on the application of implicit reference inputs to Type 1 systems.

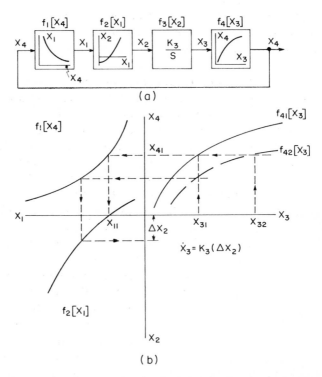

Fig. 4-15. Type 1 feedback system with nonlinear components. (a) Block diagram with implicit summing point. Note that X_2 can assume either positive or negative values, but that in the steady state, X_2 must be zero so that no further change in X_3 can take place. (b) Graphical construction for finding the steady state. With $f_{41}[X_3]$, the steady state is represented by X_{31}, X_{41}, and X_{11}, whereas with $f_{42}[X_3]$ the steady state is X_{32}, X_{41}, and X_{11}. X_2 takes on nonzero values only during transient periods when the system is reacting to a disturbance.

a. The steady-state relations in a Type 1 system are necessarily different from those in a Type 0 because of the presence of the integrating element. The block diagram in (a) depicts an example with nonlinear characteristics throughout. Note that $f_2[X_1]$ yields a zero value of X_2 for a specific value of X_1.

b. In the steady state, X_2 must be zero in order for the integrator to stop integrating. For this to be true, the several variables must take on the values X_{41}, X_{11}, and X_{31}. X_2 is zero as noted. This is the only possible steady state, because for any other values of the variables, X_3 will continue to increase or decrease as $f_3[X_2]$ continues to integrate X_2.

c. A disturbance to the system might be represented by the change in $f_4[X_3]$ from $f_{41}[\quad]$ to $f_{42}[\quad]$ as shown. Momentarily an error is developed equal to ΔX_2, and since this is positive, X_3 will increase until it reaches the magnitude X_{32}, at which point X_2 again returns to the value 0.

d. This system includes an implicit summing point in the function $f_1[X_4]$, which reverses the sign of the signal. The form of $f_2[X_1]$ allows X_2 to become zero for a nonzero value of X_1. The magnitude of X_1 providing the zero output for X_2 in effect performs the function of a reference input, and the integrating component always serves to force the system back to the state in which X_2 is zero. The regulated variable X_4 always returns to the value X_{41} after each disturbance.

e. Although there are a number of differences in the behavior of Type 0 and 1 systems, we limit the following discussion to the matter of identification. Can the two systems be distinguished from closed-loop measurements? In the Type 0 system, it has been shown that input and output variables may be made as nearly equal as desired if the loop gain is made large. In the limit a measurable difference in the magnitudes of input and output is limited by noise, and this will be true of both types of system. Thus, although a Type 1 system has an error that is ideally zero, noise prevents its attainment, and in the steady state there may well be a residual and uncorrectable error. If open-loop tests were possible, one could decide without ambiguity whether the system contained an integrating component. In the absence of a clear demonstration that such components exist biologically, we conclude that homeostats are Type 0.

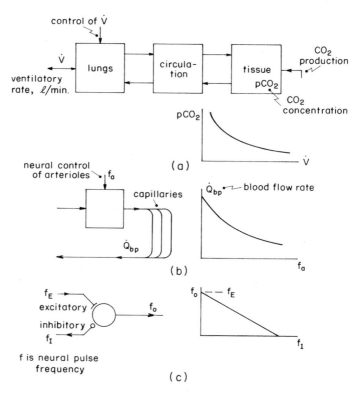

Fig. 4-16. Examples of sign reversal with biological signals. (a) In respiratory control the tissue pCO_2 is decreased as a consequence of an increase in ventilatory rate $\dot{V}$. (b) The resistance of the arterioles is under neural control such that capillary blood flow $\dot{Q}_{bp}$ decreases with an increase in neural pulse frequency f_a. (c) The inhibitory input to a neuron f_I may reduce its firing rate f_o for a constant excitatory input f_E.

a. The discussion in the previous sections has suggested that sign reversal, a necessary feature in a negative feedback regulator, may frequently be implicit within the system. If this is the case, then no explicit summing point is required. Some examples of sign reversal are described in this section.

b. The liquid level regulator in Fig. 4-12b has a control mechanism that acts upon the tank outflow. The fact that it is the outflow that is controlled provides the sign reversal needed. The sweat glands act in a similar manner, that is, increased sweat rate and evaporation following an increase in the neural signals to the glands increase the rate of heat loss. This provides the negative feedback for temperature regulation.

c. Control of a loss function is also found in respiration, where an increase in the ventilatory rate, $\dot{V}$, in the lungs decreases the concentration of CO_2 in the tissues. An increase in $\dot{V}$ decreases the average pCO_2 in the lungs with the consequence that CO_2 then diffuses at a greater rate from the cells. While there are, of course, many differences in the detailed physical processes between heat loss by evaporation and CO_2 loss by diffusion, they do have the common feature of providing a sign reversal in the appropriate signals.

d. The control of flow resistance is exemplified by the arterioles in peripheral circulation, as in (b). The muscular coat surrounding the arterioles provides a means of controlling blood flow, and since these are constrictor muscles, the flow rate varies inversely with the magnitude of muscle innervation.

e. Inhibitory signals to a neuron decrease the effectiveness of excitatory inputs, and thus provide a means of sign reversal. The relations shown in (c), while highly simplified, do depict the decrease in firing rate of a neuron that accompanies an increasing inhibitory input. This effect is only seen in the presence of an adequate excitatory drive, but may well lead to a complete cessation of firing.

f. These few examples will serve to illustrate the fact that sign reversal is readily identified in physiological systems. Furthermore, if such a process exists, there is no need for the conventional summing point representation. As a matter of fact, since sign reversal is a necessary function in a homeostat, it is frequently advisable to identify it immediately. Its location within the system is subject to considerable variation, sometimes being close to the regulated variable and in other instances relatively far removed.

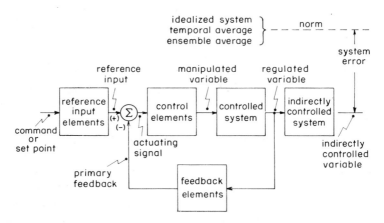

Fig. 4-17. Block diagram of a technological regulator drawn in a standard form to show the definition of various terms. The components shown here may, or may not, be found in a given regulating system. In many instances the biological regulating system is better represented by the block diagram of Fig. 4-3 or 8-8. The actuating signal and manipulated variable cannot always be distinguished from each other.

a. This diagram is an attempt to develop a perfectly general description of a technological feedback system for engineering purposes, but such a generalized representation is soon found to be inadequate. In the broader area of biological regulating systems, the limitations are readily apparent. However, in spite of the inadequacies, it is possible to use such a diagram effectively to discuss certain properties of interest from both engineering and physiological viewpoints.

b. Components outside the feedback loop (reference input elements, indirectly controlled system) do not receive the benefits of feedback and the sensitivities of these components are not reduced. The indirectly controlled system is specifically shown to illustrate the fact that the variable which one would like to regulate (here denoted the indirectly controlled variable) is frequently not accessible to measurement, and one must use some related quantity.

c. The separation of the components into control elements and controlled system is quite arbitrary in many instances. From an engineering viewpoint, the controlled system is that set of processes which is given, and over which some control is to be exercised. The control elements include those components which must be added by the designer to effect regulation and for this reason are readily identified. In a physiological system, any such distinction is usually arbitrary and may well be meaningless.

d. The previous discussion of error or unbalance in connection with Fig. 4-11 may be further amplified. The idealized system is a concept sometimes introduced into engineering discussions to permit a more precise definition of error. It is a physical system that produces the desired norm, and is idealized to the extent that it is unaffected by all the environmental factors that disturb the actual system. With a norm defined in this manner, it is possible to define an error as shown.

e. To extend this discussion to physiological regulators, it is reasonable to introduce both a *temporal* and an *ensemble norm*. The temporal norm might be an average over time on one individual, with the averaging period taken over a time interval sufficiently long to smooth out daily variations. For that one individual an "error" with respect to his temporal norm would seem to have significance. Similarly, an average taken over a population would yield an ensemble norm, and a similarly defined "error."

f. It is important to recognize that the error as defined in any of these ways is purely conceptual, and is not a physical quantity or variable within the system that acts as a signal in the feedback loop.

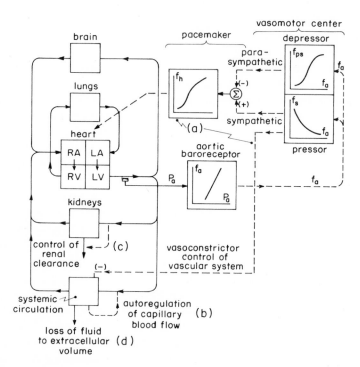

Fig. 4-18. Overall view of the circulatory system with some of its regulatory processes. (a) Regulation of systemic arterial pressure in which peripheral systemic resistance and heart rate are the manipulated variables. (b) Autoregulation of capillary flow by direct constriction or dilitation of the vessels in response to metabolites in the blood. (c) Total fluid volume in the body is controlled by excretion and reabsorption in the kidneys. (d) Blood volume changes resulting from diffusion of fluid from the circulatory system into the extracellular space.

a. The circulatory system performs a variety of functions associated with the flow of nutrients, waste products, and heat throughout the body. Since the demands of these separate functions vary from one region to another, and also with the loads placed upon the organism, it is necessary that a number of controls and regulatory systems be available. This figure is designed to show the blood pressure regulator, or baroreceptor reflex (a), in the context of some of the other regulating mechanisms (Guyton, 1971).

b. The feedback regulation of arterial pressure P_a serves to maintain this pressure throughout the body at a relatively constant value, so that the other regulators will not have to contend with changes in blood pressure as a disturbance. The baroreceptor reflex has a response time measured in seconds, so that it can react rapidly to changes in posture, but on the other hand it adapts so that other, longer-lasting processes (such as control of blood volume) are required for regulation over longer periods.

c. The baroreceptors located in the aortas and a number of other points within the body send signals proportional to pressure f_a to the medullary or vasomotor center. There are actually three manipulated variables: (1) vasoconstriction, (2) heart rate, and (3) heart contractility (not shown).

d. Not all regulatory processes make use of feedback as was suggested in Section 1.2. For example, increase in pressure in the capillary bed will increase the movement of fluid into the extracellular space as shown in (d). This does not involve feedback, and is better termed *compensation.*

e. Another process acting at the capillary level adjusts the blood flow to match the need of local tissue for oxygen. This is frequently termed *autoregulation,* as in (b). It appears that the local concentration of O_2 and other metabolites has a direct effect upon the capillary musculature, and the process acts as a feedback regulator for O_2 (Burton, 1965).

f. The present evidence suggests that the control of peripheral resistance occurs entirely by means of vasoconstrictor fibers, and the vasodilator fibers are not active in blood pressure regulation (Mountcastle, 1968, p. 188). This means that with normal pressure, there is a certain amount of vasoconstrictor tone, and that either an increase or a decrease in the constriction is possible.

g. Since a large number of regulatory processes are associated with circulation, considerable interaction might be expected between them. Fortunately, the response times of these homeostats are fairly well separated so that to some extent one may consider one regulator at a time.

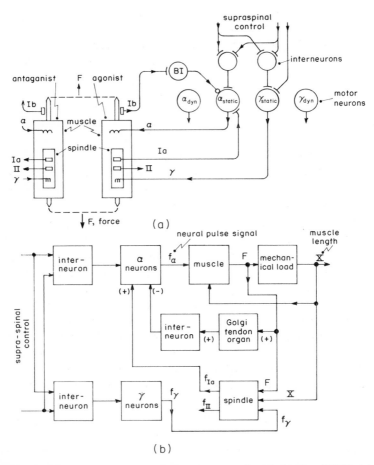

Fig. 4-19. Neuromuscular control and the monosynaptic spinal reflex (MSR). (a) Quasi-anatomical diagram showing an antagonistic muscle pair acting at one joint together with some of the neural pathways for the agonist muscle. (b) Block diagram of the MSR for one muscle. f designates neural pulse signal, with the subscript identifying the pathway.

a. This reflex, also termed the monosynaptic spinal reflex (MSR), is a feedback system that forms a basic unit in neuromuscular control of posture and movement. Some of the components and neural circuits for one muscle are shown in (a); those for the other muscle are similar but not necessarily identical (Granit, 1970).

b. The muscles contract in response to signals in the α (efferent) nerves, which terminate in motor endplates on the extrafusal muscle fibers.

c. The proprioceptive units (spindles and tendon organs) send neural (afferent) signals that report the state of the muscle to the spinal cord. The Ia and II fibers signal the muscle length, and in part, the rate of change in length. The Ib fibers send a signal representing muscle tension. Thus, two feedback paths exist, one for muscle length, and one for tension.

d. The spindle itself contains intrafusal muscle fibers, the tension of which is controlled by the γ fibers. These signals serve to "recalibrate" the spindle.

e. The block diagram depicts some of the functional relations in these feedback systems, but only for one of the muscles. It is important to note that with each muscle, there are hundreds or thousands of α and γ motor neurons and likewise a large number of each of the types of proprioceptive units. Thus, a complete diagram for a single muscle possibly could contain thousands of paths similar to the one shown, acting more or less in parallel.

f. The α as well as the γ motor neurons fall into two somewhat separate classes associated with static (tonic) and dynamic (phasic) behavior. Thus, there appears to be some specialization of function among the populations of motor neurons, although the precise role of each is not completely clear.

g. Sign reversal for negative feedback is inherent in the circuits shown. An increase in muscle length will cause an increase in the Ia and α signals, and thus greater muscular contraction. An increase in tension is sensed by the tendon organs and brings about a reduction in the contraction.

h. The precise roles played by the α and γ motor neurons in the execution of specific movements is not clear, but probably varies with the character of the movement, and is adjusted by supraspinal control. Although the desired state for the muscle is in a sense established by the γ motor neurons, in some instances the two groups of neurons act in sequence, whereas in other situations they act in concert. In addition, the supraspinal centers can modify both the static and dynamic properties of these neurons.

1. Homeostatic mechanisms have been acquired by organisms as a consequence of evolutionary pressures, and have become such a fundamental aspect of biological systems that it is difficult to discuss a process apart from its associated feedback. On the other hand, in technological systems the feedback is clearly the addition of man, and for the most part is readily distinguishable from the process it regulates.

2. The consequences of a disturbance to a set of processes having no feedback can frequently be identified by inspection, and are usually confined to only a portion of the system. The addition of feedback to the system results in all variables being coupled in a cyclic manner, and in general all variables are affected by any disturbance.

3. The steady-state operating point is given by the simultaneous solution of all the functions describing the steady state. The fact that all the variables are related in a cyclic order leads to a set of operational rules for identifying a negative feedback system as one having (a) this cyclic coupling, (b) a sign reversal, and (c) a unilateral loop transmission.

4. A disturbance to a homeostatic system will result in a change in one or more of the steady-state characteristics, and a new operating point. The perturbations resulting from a given disturbance will change all the system variables, to varying degrees and in varying directions.

5. Loop gain is the amplification that a small constant signal undergoes after one transit around the feedback loop. In contrast, the closed-loop gain is the transmission from a specific input variable (the reference input) to the regulated variable with the loop closed. The closed-loop gain is a measure of how closely the regulated variable equals the reference input.

6. The control mechanism for a feedback regulator usually performs two functions. It reverses the sign of the feedback signal, and introduces a constant term (the reference input). These two functions are frequently interrelated so that for a specific control mechanism, an analysis of its operation is required to show what factors affect each of these two functions.

7. The loop gain may be calculated by multiplication of the gains of each of the components in the loop. Although this product is negative, the loop gain is usually given as the absolute value of the product, and is positive. The product with the negative sign retained has been termed the homeostatic index.

8. Regardless of the precise location of a disturbance, negative feedback has the effect of reducing the perturbation in the regulated variable, compared to what it would be in the absence of feedback.

9. The system sensitivity is defined as the per unit change in the regulated variable divided by the per unit change in the disturbed quantity. Thus a good regulator has low sensitivity, and feedback serves to decrease the sensitivity. An incremental sensitivity is defined for small but finite disturbances, and a differential sensitivity for infinitesimal changes.

10. The sensitivity is reduced by an increase in loop gain, and at all gains the incremental sensitivity is greater than the differential sensitivity.

11. Systems having nonlinear characteristics can be approximated by straight lines tangent to the steady-state characteristic at the operating point. Such linear approximations are valid for small disturbances, but clearly show that a disturbance may change the loop gain as well as introduce bias terms at various points in the system.

12. In a unity feedback system with large gain, the operating point has a value such that the regulated variable is nearly equal to the reference input. The difference between these two quantities is termed the actuating signal. This difference is sometimes designated as an "error" but this usage has little significance when components are present in the feedback path.

13. In the control of body temperature by sweating, it is the heat loss that is the manipulated variable, and the sign reversal is inherent in the sweating process. Although conservation of energy requires an equality of heat inputs and outputs in the steady state, this does not assure the same temperature in all steady states.

14. The thyroid–pituitary system is a feedback regulator in which thyroxine is the regulated variable, and thyroid stimulating hormone is the manipulated quantity. With respect to the whole organism, this system is a minor feedback loop that appears in all homeostats in which thyroxine plays a role.

15. When an integrating element is introduced into a feedback regulator, there results a system that will (ideally) reduce the actuating signal (error) to zero (a Type 1 system). A servomotor is a good approximation to an ideal integrating element.

16. Analysis shows that in the steady state, the integrator must have zero input, and a Type 1 system is thus "astatic" in that it is able to maintain any given magnitude of the regulated quantity with a zero signal at the integrator input.

17. Sign reversal of a biological signal may occur by (a) a loss process as in the case of the lungs and body tissue where an increase in ventilatory rate lowers the CO_2 concentration, (b) a controllable resistance such as the arterioles in which capillary blood flow is reduced with increase in neural signal, and (c) neural inhibition, which can reduce the firing rate of a nerve cell.

4.21   EPITOME   125

18. Although a generalized block diagram serves to define a number of terms, such a representation can only show a fairly common arrangement and cannot describe all possible cases. It is to be expected that biological regulators will show wide variations from any standard form.

19. Among the many regulators associated with the circulatory system, the baroreceptor reflex is one of the fastest. It acts by control of peripheral resistance, heart rate, and heart contractility.

20. The stretch reflex, associated with practically all skeletal muscles, provides regulation of muscle length and muscle force. The static and dynamic properties of this feedback regulator may be adjusted by means of signals from supraspinal centers.

Problems

1. In Fig. 4-1, evaluate the effect of the disturbance $+\Delta R_4$ *with the feedback control intact.* Can one find the sign of the incremental steady-state changes (as in the table) by inspection? Carry out such a calculation by a construction similar to Fig. 4-3.

2. The loop gain, K_1K_2, in Fig. 4-4 can conceivably be divided into component gains K_1 and K_2 in any manner. Assume that $K_1K_2 = 100$. Find $\Delta X_2/\Delta X_3$ for $K_2 < 1$ and $K_2 > 1$.

3. The respiratory chemostat may be considered as a regulator of tissue CO_2 in the body and represented most simply by the block diagram in Fig. P4-3. See also Fig. 8-7a.

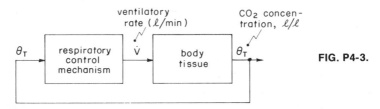

FIG. P4-3.

(θ_T) tissue CO_2 concentration (liters/liter); $(\dot{V})$ pulmonary ventilation (liters/min).

In the steady state, resting and breathing normal air, these variables are related by the functions:

$$\dot{V} = 471\,\theta_T - 246; \qquad \theta_T = 0.365 + 0.85/\dot{V}.$$

(a) Find the magnitude of these two variables at the operating point. Estimate the open-loop gain. Compare algebraic and graphical methods for finding these quantities.

4 INTRODUCTION TO FEEDBACK; THE STEADY STATE

(b) When breathing a 5% CO_2 mixture, the second expression becomes

$$\theta_T = 0.525 + 0.85/\dot{V}.$$

Under these conditions find the operating point and the loop gain.

4. Confirm one point on each curve in Fig. 4-9. For each case, sketch the block diagram using (a) the undisturbed magnitudes and (b) the disturbed magnitudes of K and X_3.

5. In the liquid level regulator of Fig. 4-12, the tank may be said to "integrate" the difference $\dot{Q}_i - \dot{Q}_0$. In what sense, if any, may this be said to be a Type 1 system? Consider all disturbances $\Delta\dot{Q}_i$, ΔK, ΔY_0.

6. The following questions refer to the system of Fig. 4-15:

(a) What effect (if any) does the constant K_3 have upon the steady state of the system, or on other aspects of its performance?

(b) Suppose that function $f_1[X_4]$ is disturbed. How does this affect the steady state? Do the same for $f_2[X_1]$.

(c) Make a construction as in Fig. 4-14b, showing a similar disturbance, except with an initial effect of producing a *negative* value of ΔX_2.

(d) What aspect of the function $f_2[X_1]$ makes it quite unlikely that a biological example could be found to have this characteristic?

7. Suggest another example of a biological process that produces a reversal in the sign of a signal. How is sign reversal accomplished in the following cases?

(a) Temperature regulation by sweating.

(b) Temperature regulation by shivering.

(c) Temperature regulation by vasomotor control.

8. Make a table of the regulatory processes associated with the circulatory system, starting with those shown in Fig. 4-18 and adding as many more as possible. For each process, indicate (1) the regulated variable, (2) the manipulated variable, (3) whether it involves feedback or not, and (4) an approximate measure of its response time. For further references, see Guyton (1971), Mountcastle (1968), and Burton (1965).

9. With regard to Fig. 4-5 and the process in Fig. 4-1:

(a) Verify the relations for $\dot{Q}_1$ and $\dot{Q}_{1m}$ in Fig. 4-5.

(b) Show how the operating point changes if the length L is increased as in Fig. 4-5b. Is the change in Y_2 equal to the change in L? Explain.

(c) The ratio a/b is to be increased as in Fig. 4-5c, but the original operating point is to be retained. Show how this may be accomplished.

10. Discuss the positional control of some portion of the body (head, eye, arm) with regard to the Type 1 and Type 0 systems discussed in this

chapter. Would you expect to find an error in the steady-state position? Which type of system seems to best describe the physiological system? Might visual feedback be a factor?

11. The Type 1 system of Fig. 4-14 has a zero error (ideally) when the input θ_i is a constant. Will there be an error (ϵ nonzero) if the input is a constant velocity, that is, $\theta_i = at$? Discuss this case from a purely physical standpoint.

References

Adolph, E. F. (1943). "Physiological Regulations." Jacques Cattell Press, Lancaster, Pennsylvania.

Adolph, E. F. (1968). "Origins of Physiological Regulations." Academic Press, New York.

Brown-Grant, K. (1969). Current Views on the Behavior of the Thyroid–Pituitary System, in "Hormonal Control Systems" (E. B. Stear and A. H. Kadish, eds.), Suppl. 1, Mathematical Biosciences. Amer. Elsevier, New York.

Burton, A. C. (1965). "Physiology and Biophysics of the Circulation." Yearbook Publ., Chicago, Illinois.

Distefano, J. J., and Stear, E. B. (1969). Modeling and control aspects of thyroid function, in "Hormonal Control Systems" (E. B. Stear and A. H. Kadish, eds.), Suppl. 1, Mathematical Biosciences. Amer. Elsevier, New York.

Granit, R. (1970). "The Basis of Motor Control." Academic Press, New York.

Guyton, A. C. (1971). "Textbook of Medical Physiology," 4th ed. Saunders, Philadelphia, Pennsylvania.

Horowitz, I. M. (1963). "Synthesis of Feedback Systems." Academic Press, New York.

Jones, R. W. (1951). Effects of loads and disturbances upon feedback controller. *Trans. AIEE* **70**, 460–464.

Jones, R. W. (1960). Some properties of physiological regulators. *Proc. Int. Conf. Fed. of Auto. Control, 1st, Moscow,* pp. 586–590.

Jones, R. W. (1969). Biological control mechanisms. *In* "Biological Engineering" (H. P. Schwan, ed.). McGraw-Hill, New York.

Langley, L. L. (1965). "Homeostasis." Van Nostrand-Reinhold, Princeton, New Jersey.

Mountcastle, V. B. (1968). "Medical Physiology," 12th ed., Vols. I & II. Mosby, St. Louis, Missouri.

Purves, H. D. (1964). Control of thyroid function, in "The Thyroid Gland" (R. Pitt-Rivers and W. R. Trotter, eds.), Vol. 1, pp. 1–38. Butterworths, London.

Riggs, D. S. (1963). "Mathematical Approach to Physiological Problems." Williams & Wilkins, Baltimore, Maryland.

Yamamoto, W. S. (1965). Homeostasis, continuity and feedback, in "Physiological Controls and Regulations" (W. S. Yamomoto and J. R. Brobeck, eds.). Saunders, Philadelphia, Pennsylvania.

Feedback Systems; Dynamic Behavior || *Chapter 5*

In the previous chapter it was shown that negative feedback is able to reduce the effects of disturbances from the environment, and thus to decrease the system sensitivity. In this sense, the addition of feedback is said to produce a *regulated,* or *homeostatic, system.* These two terms will be used interchangeably, and will denote a system for which negative feedback pathways can be identified.

Although the steady-state consequences of feedback are most significant, these are not the only effects, and we turn now to a consideration of the dynamic properties. In the open loop, a group of components, when connected in sequence, will exhibit dynamic properties that are directly related to those of the individual components. However, with feedback added, the dynamic properties may change radically and the relation between the dynamic behavior of the system and that of the individual components may become quite subtle. In technological applications, this effect upon system dynamics may be the principal reason for using feedback. Whether this is also the case for physiological processes remains to be seen (Machin, 1964).

Homeostats are being disturbed continually, and rather than reaching a pure steady state, they undergo a continuing sequence of transient responses to these disturbances. Hence, an understanding of dynamic properties, in both the steady state and following some imposed disturbance, is fundamental to a study of feedback system behavior.

The disturbances considered in this chapter are, for the most part, step function changes in some parameter or input quantity. In actuality, almost any temporal form of disturbance may be encountered, and the step function is chosen solely for its simplicity. The response to other forms of disturbance requires considerably more in the way of calculations than can be undertaken here.

At a number of points we take the view that feedback can be readily added to a set of components, and thus one is justified in discussing both the open-loop and closed-loop behaviors of the same system. While such a comparison may be quite feasible in a technological system, it may not be at all possible in a biological context. Despite the practical limitations, it is still conceptually possible to contrast open- and closed-loop performance, and this is done frequently in the following pages. This somewhat artificial viewpoint is adopted so as to clarify the consequences of feedback per se (Eckman, 1966).

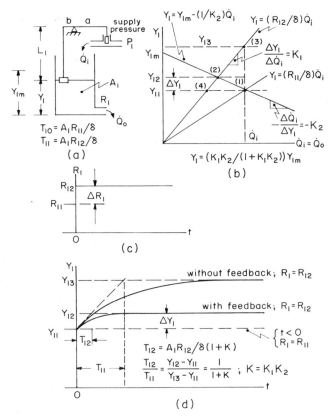

Fig. 5-1. Changes in process behavior by virtue of feedback. (a) Flow process with feedback regulation of liquid level, subjected to a change in outflow resistance R_1 as shown in (c). (b) Steady-state characteristics showing the operating point prior to the disturbance (1), and after the disturbance (2). If there were no feedback, and $\dot{Q}_i$ remained constant, the new operating point would be (3). (d) Responses of Y_1 both with and without feedback.

a. This section pertains to the modifications in system behavior brought about by feedback. The system shown in the figure serves to regulate the liquid level Y_1 and is of the first order.

b. The disturbance is a step function change in R_1, denoted ΔR_1 in (c). The choice of R_1 as the disturbed variable, and the step function change, were made solely for clarity of exposition.

c. Part (b) of the figure shows the steady-state characteristics for both the flow process and the control linkage. Prior to the disturbance, and with $R_1 = R_{11}$, the operating point is (1), whereas after R_1 becomes R_{12} the operating point moves to (2) if the feedback is effective. On the other hand, if the feedback was not acting, the operating point would be (3). Feedback has served to reduce the deviation in Y_1 that occurs with this disturbance.

d. The dynamic behavior shown in (d) is described by a single exponential component. However, feedback has reduced the time constant in this case to about one third of its original value. What brings this about?

e. At the first instant, following the increase in R_1, there is an immediate decrease in $\dot{Q}_0$, and Y_1 starts to rise at a rate that is the same for both feedback and nonfeedback cases because the feedback mechanism is not effective at this first instant. However, as soon as Y_1 has increased by even a small increment, $\dot{Q}_i$ is caused to decrease, and $\dot{Q}_0$ starts to recover from the sudden decrease it experienced at the start. This behavior of $\dot{Q}_0$, although not shown in this figure, is readily sketched (Problem 5-1). Inasmuch as the final value of Y_1 with feedback (Y_{12}) is less than the value it would reach in the absence of feedback (that is, Y_{13}) the transient period is correspondingly shortened, and the time constant reduced as shown.

f. The perturbation in the steady state and the time constant are both reduced by the same factor, $1/(1 + K)$, where K is the gain in the disturbed state. This may be demonstrated by writing the appropriate differential equation (see Section 5.3 and Problem 5-1).

g. In this example, the dynamic responses with and without feedback are of the same form, although for higher order systems the entire character of the dynamic behavior may change drastically. In a first order system, feedback serves to reduce the effective time constant, but in general the effect of feedback is to increase the speed of response to imposed disturbances.

h. In this example, one should distinguish between three different time constants. Prior to the disturbance, $T_1 = T_{10}$, but this was of no interest here. Following the disturbance, $T_1 = T_{11}$ *if the feedback was absent,* and T_{12} with R_{12} and the feedback effective.

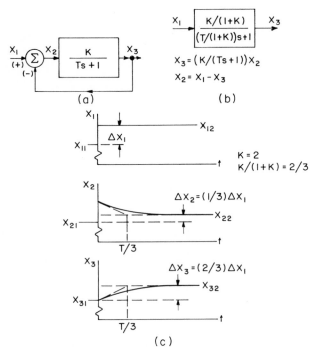

Fig. 5-2. Comparison of closed-loop and open-loop behavior. (a) Conventional block diagram of a first-order feedback system. (b) Closed-loop transfer function of the same system shown as an equivalent open-loop block diagram. (c) Response of the closed-loop system to a step function increase in the input signal X_1.

5   FEEDBACK SYSTEMS; DYNAMIC BEHAVIOR

a. The previous example of a first-order liquid level regulator enables one to visualize the physical changes that take place in a feedback system following a disturbance. To generalize the discussion, the present figure depicts a similar, but not identical, system by means of its transfer function and block diagram (see Fig. 3-13). The present figure may be considered as representing a large number of similar first-order systems, although the validity of the transfer function must be established in each instance, either by physical reasoning or directly from experimental observations.

b. In this figure, $X_2 = X_1 - X_3$, whereas the corresponding relation in Fig. 5-1 is $\dot{Q}_i = K_i(Y_{1m} - Y_1)$. With this modification, the following identifications may be made between the two systems: $X_1 \rightarrow Y_{1m}$, $X_2 \rightarrow Y_{1m} - Y_1$, $X_3 \rightarrow Y_1$.

c. The block diagram in (a) is the conventional representation of a single-loop feedback system, while the one in (b) is termed the equivalent open-loop system. The transfer function in (b) is readily obtained from the system equations shown. Comparison of the two transfer functions shows that the gain (numerator) is reduced by the factor $1/(1 + K)$, and the time constant is also reduced by the same factor. The responses sketched in (c) have been calculated for an assumed value of $K = 2$.

d. The behavior of a given system depends not only on the transfer function, and the character and location of the disturbance, but also upon the initial conditions in the system before the disturbance occurs. In this and the previous figure, it has been assumed that the process was initially at rest with all variables at some constant values. However, in Fig. 5-1, the disturbance was an increase in R_1, whereas here it is an increase in X_1.

e. Note that all dependent variables, as sketched in (c), respond with the same time constant even though the transients may have different signs.

f. It will be noted that the open- and closed-loop transfer functions in (a) and (b) have identical forms. Thus, from measurements made at the two terminals X_1 and X_3, it would be impossible to determine whether or not feedback was actually present. One concludes that although feedback may drastically alter the behavior of a physical system, the properties remain those of a system of the same general type, and no clues are provided the external observer that would enable him to identify feedback. Some important exceptions to this are discussed subsequently.

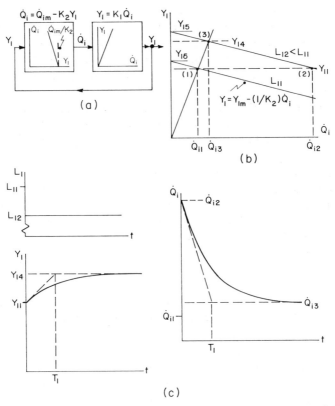

Fig. 5-3. Change in the operating point of a feedback system without a change in loop gain. (a) Block diagram of the liquid level regulator in Fig. 5-1. (b) Steady-state characteristics and operating points for a decrease in the length of L_1. Note that the axes in this diagram are interchanged compared to the ones used in Fig. 4-5. This has no effect on the results or conclusions. (c) Dynamic response to the disturbance in L_1. These curves are drawn to approximately the same scale as in (b).

     5   FEEDBACK SYSTEMS; DYNAMIC BEHAVIOR

First-Order Systems; Change in the 5.4
Operating Point: Fig. 5-3

a. This figure portrays the same system as that in Fig. 5-1, but under the condition of a change in the operating point brought about by a decrease in the length of link L_1. Although a step function change in the length L_1 may not seem to be too realistic, disturbances of this character are frequently encountered in a variety of physical systems.

b. Part (b) of the figure is drawn for a loop gain of about nine. The initial operating point is at (1), but it jumps to (2) immediately following the step function disturbance. The operating point then moves in an exponential manner to (3), and this becomes the final steady-state operating point.

c. The response curves in (c) have been drawn to the same scale as that used in part (b) so that the relative magnitudes of the transients in Y_1 and $\dot{Q}_i$ can be compared. Y_1 increases exponentially with time as it moves from its initial value Y_{11} to its final value Y_{14}. On the other hand, $\dot{Q}_i$ exhibits a sudden increase from $\dot{Q}_{i1}$ to $\dot{Q}_{i2}$, and then decreases to its final value $\dot{Q}_{i3}$ exponentially. The transient portion of each of these responses is characterized by the same time constant T_1.

d. The curves in this figure illustrate responses that are continuous as well as discontinuous. A *continuous function* is one that exhibits only infinitesimal changes in the dependent variable for similar changes in the independent quantity. In contrast, a *discontinuous function* is one that exhibits a finite (not infinitesimal) change in the dependent variable for an infinitesimal change in the independent variable. $\dot{Q}_i$ is discontinuous at $t = 0$, where it suddenly jumps from one value to another. On the other hand, Y_1 is continuous. However, dY_1/dt is discontinuous at $t=0$, where it suddenly changes from 0 to a positive value.

e. On physical grounds, Y_1 cannot undergo a discontinuous change because to do so would require the transport of a finite amount of liquid in zero time. The fact that $\dot{Q}_i$ does change in a discontinuous manner is permitted by the assumption that the value-operating mechanism contains no dynamic parameters, and may thus change its position instantly. Furthermore, the liquid is assumed to have negligible inertia so that its flow rate can change suddenly.

f. From this discussion, it should be concluded that, in general, physical variables cannot undergo discontinuous changes. Rapid changes that approach discontinuities are possible, but only when the pertinent dynamic parameters are so small as to be negligible compared with others.

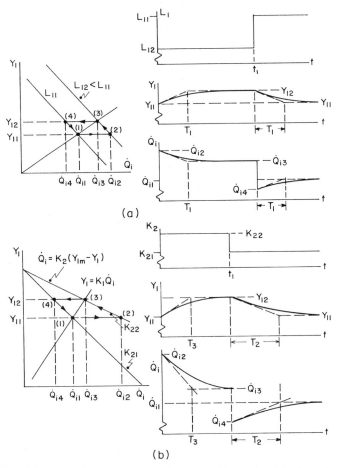

Fig. 5-4. The effects of disturbances to a feedback system showing the trajectory of the representative point during the transient period. (a) Change in the set point produced by a decrease in the length L_1. (b) Change in the loop gain produced by an increase in the lever arm ratio a/b in Fig. 5-1.

a. This figure depicts the behavior of the physical system appearing in Fig. 5-1 following two different disturbances. The perturbation caused by a reduction in the length of link L_1, as shown in (a), is contrasted with the effect produced by a change in the fulcrum position in (b). In the latter instance, the fulcrum is shifted to the left, a change that produces an increase in loop gain K_1.

b. The reduction in the length L_1 merely shifts the steady-state characteristic parallel to itself as already shown in Figs. 4-5 and 5-3. The resulting transient involves an increase in $\dot{Q}_i$ from $\dot{Q}_{i1}$ to $\dot{Q}_{i2}$, followed by a decrease to $\dot{Q}_{i3}$. The first portion of the transient is discontinuous following the argument of the previous section, whereas the second portion is expontential with the time constant T_1.

c. The path (1–2–3) along which the system variables move is termed a *system trajectory,* and points lying on this trajectory represent the values of the system variables (Y_1 and $\dot{Q}_i$) during the transient process. Upon the return of L_1 to its original length, the trajectory followed is (3–4–1). The response time constant, which is determined by the compartment size, the outflow resistance, and the loop gain, is not affected by this disturbance and therefore the response has the same exponential term for both the increase and decrease in L_1.

d. In part (b) of the figure, the disturbance is an increase in gain K_1, and the resulting trajectories are again marked 1–2–3 and 3–4–1. In this instance, however, the time constant is reduced by the increase in K_1 so that $T_3 < T_2$.

e. The system trajectories, also called *phase trajectories,* provide a view of system behavior that is quite different from that given by the response curves in the time domain. Time does not appear explicitly in these diagrams but is an implicit variable, and a value for t can be attached to each point along the trajectory. From the assumptions underlying this figure, points (1) and (2) represent the conditions at zero time, and the transition from (1) to (2) occurs instantly. The final steady state represented by point (3) is reached in infinite time in an exponential manner. The point moving along a trajectory to describe the dynamic behavior of a system is termed the *representative point.* The term operating point will be reserved for the steady-state points on the trajectories.

f. The plane on which these trajectories are drawn is termed the *phase plane,* the usefulness of which becomes greater if nonlinearities are present (Chapter 9).

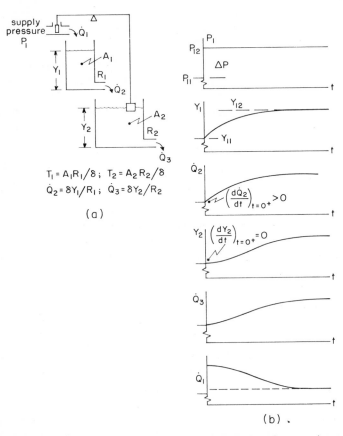

$$T_1 = A_1 R_1 / \delta \; ; \quad T_2 = A_2 R_2 / \delta$$
$$\dot{Q}_2 = \delta Y_1 / R_1 \; ; \quad \dot{Q}_3 = \delta Y_2 / R_2$$

(a)

(b) .

Fig. 5-5. Second-order fluid level regulating system consisting of two noninteracting compartments. (a) Regulated system. (b) The system is disturbed by a step function increase in the supply pressure P_1. The subsequent response of all the dependent variables is shown.

a. This system consists of two noninteracting compartments provided with feedback regulation of the level Y_2. The disturbance is an increase in supply pressure P_1, which increases the magnitude of $\dot{Q}_1$ for a given valve position. Following the argument of Chapter 2, it will be found that all the system variables approach steady-state values that are somewhat greater than prior to the disturbance.

b. All of the response curves contain the same two exponential transients, but these are combined in different ways to produce the variety of shapes shown. The time constants of the exponential components are not the same as the compartment time constants, T_1 and T_2, but are related to them in a manner somewhat similar to that discussed for the first-order system in Section 5.2. In general, the response is faster with the feedback present.

c. The step function disturbance occurring at the input results in a prompt (but continuous) response of Y_1 and $\dot{Q}_2$. Although these two curves may appear to be single exponentials, they are the sum of two exponential curves, as described above. The two curves Y_1 and $\dot{Q}_2$ are identical except for a scale factor.

d. The response curves for Y_2 and $\dot{Q}_3$ show no immediate change following the disturbance because there is no effect upon the second compartment until $\dot{Q}_2$ has changed by a finite amount. This is expressed mathematically by stating that the derivative dY_2/dt is zero immediately following the disturbance, that is, at $t = 0+$. Since $\dot{Q}_3$ is proportional to Y_2, the same is true of $\dot{Q}_3$, and its derivative is initially zero.

e. $\dot{Q}_1$ is related to Y_2 in a linear manner, as discussed previously in connection with the feedback mechanism. Thus, these two curves are the same except for a change in sign and a change in scale. The disturbing signal is not propagated through the entire feedback loop until $\dot{Q}_1$ starts to fall.

f. At $t = 0$ there is a step function change in $\dot{Q}_1$ produced by the disturbance in P_1. This assumes that both of these variables can indeed change in such a discontinuous manner. From a practical standpoint, if the changes in P_1 and $\dot{Q}_1$ are very fast with respect to the other changes in the system, then such a discontinuity is a good approximation to the actual performance.

g. The calculation of the response time constants for this second-order system requires the solution of an algebraic equation of the second degree, namely, the characteristic equation as described in Section 3-16.

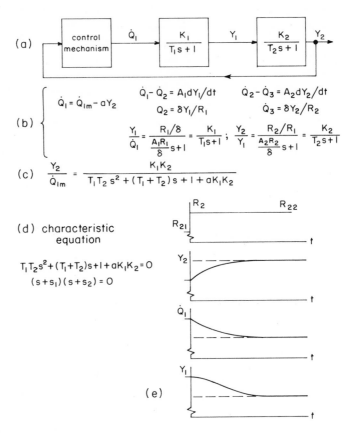

Fig. 5-6. (a) Block diagram for the feedback system of Fig. 5-5. (b) Equations for the components and their transfer functions. Note that the equation for the control mechanism is not identical with that shown in Fig. 5-1, but the two are entirely equivalent. (c) Closed-loop transfer function, and (d) the characteristic equation. (e) Responses of the dependent variables following a step function increase in R_2.

5 FEEDBACK SYSTEMS; DYNAMIC BEHAVIOR

a. This figure depicts the block diagram and transfer functions for the same physical system shown in Fig. 5-5. The control mechanism is represented by the relation $\dot{Q}_1 = \dot{Q}_{1m} - aY_2$, which could also have been shown in the block diagram in the manner used in Fig. 4-6d.

b. The relations given in (b) have been obtained from the conservation of mass and loss equations for each compartment. The transfer function follows directly. In (c), these transfer functions have been used to form the closed-loop transfer function whose denominator is the characteristic equation of the system. Note that the derivative d/dt and the Laplace variable s have been used pretty much interchangeably in these relations.

c. The characteristic equation (d) is an algebraic equation having two roots, $-s_1$ and $-s_2$, and the character of these roots will determine the nature of the response, as previously described in Section 3.16. For this figure, the two roots have been assumed to be real and negative so that the dynamic response contains two decaying exponential terms.

d. The disturbance under consideration is a step function increase in R_2 as shown in (e). Examination of the transfer function for the second compartment will show that this disturbance affects both its numerator as well as the time constant. The fact that the numerator is changed means that there is a new value for the steady-state gain, and the operating point following the disturbance is different from what it had been previously.

e. The dynamic behavior of some of the variables is shown in (e). None of these quantities exhibit a discontinuous change immediately following the disturbance, although examination of the physical system will show that $\dot{Q}_3$ does change in such a manner. Both Y_2 and $\dot{Q}_1$ start to change immediately, but do so in a continuous fashion. In contrast, Y_1 shows no immediate change, and is not affected until there is a net outflow in the first compartment.

f. As mentioned in a previous section, the dynamic response of each of the dependent variables contains the same two exponential terms, but although their time constants are the same, the signs of the terms and their relative magnitudes may be different. This accounts for the different shapes of these curves.

g. The addition of feedback to these two compartments has not changed the number of modes of free vibration, but it generally does modify the character of these modes. With some other choice of parameters, the two modes could well be of a damped oscillatory character.

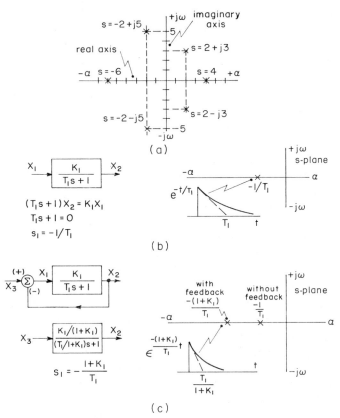

Fig. 5-7. (a) Two pairs of conjugate complex numbers, $(-2 \pm j5)$ and $(2 \pm j3)$, together with two real numbers, -6 and 4, are shown plotted on the s-plane. (b) The transient mode associated with a real root of the characteristic equation $(-1/T_1)$. The root is also termed a pole of the transfer function. (c) The addition of feedback has the effect of decreasing the time constant of a first-order system. This is equivalent to increasing the magnitude of the root of the characteristic equation.

5   FEEDBACK SYSTEMS; DYNAMIC BEHAVIOR

a. Although the previous sections have described the dynamic behavior of first- and second-order systems in the time domain, it becomes difficult to extend that discussion to systems of higher order without extensive mathematical development. At this point we introduce another conceptual tool, the *s*-plane, which helps one to visualize the behavior of more complex systems (Milsum, 1966).

b. It was suggested earlier that the dynamic modes of behavior were found from the roots of an algebraic equation (the characteristic equation), and that these roots were either real numbers or pairs of conjugate complex numbers. The *s*-plane is the medium for plotting these roots. A complex number such as $s = \alpha + j\omega$ consists of two portions, a real part α and a so-called imaginary part $j\omega$. The j may be considered as a tag to identify the second component, the magnitude of which is ω. Some examples are shown plotted on the *s*-plane in (a).

c. The first-order process shown in (b) will exhibit a single exponential mode in its response having a time constant of T_1. At this point, we are not concerned with the magnitude nor the sign of this component, but only with the time constant. The root of the characteristic equation, $-1/T_1$, is plotted on the *s*-plane, and we identify this root with the exponential mode shown. Since the characteristic equation is identical with the denominator of the transfer function, the root of the denominator is said to be a *pole* of the transfer function, a term borrowed from complex function theory.

d. The presence of feedback, as in (c), changes the denominator of the equivalent transfer function, and its root (the pole) now becomes $-(1 + K_1)/T_1$ and appears at the corresponding point on the negative real axis. The exponential mode associated with this pole has a shorter time constant than in the open loop, and the mode (transient) disappears in less time. Exponential modes associated with poles lying at some distance from the origin thus have short time constants, and the poles close to the origin give rise to transients that persist for extended periods of time.

e. Although this mode of representation is equally applicable to the open loop or the closed loop, the location of the poles is quite different in the two cases. The poles must of course be the roots of the characteristic equation that applies to the case under study. Although the cluster of poles on the *s*-plane does not provide a complete picture of dynamic behavior, it does indicate the dominant poles (those close to the origin), and in contrast, those that lie well removed from the origin and hence may be negligible.

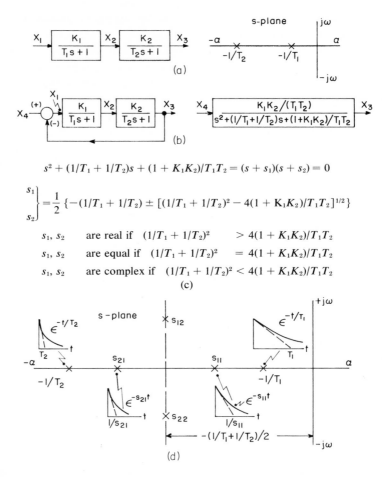

$$s^2 + (1/T_1 + 1/T_2)s + (1 + K_1K_2)/T_1T_2 = (s + s_1)(s + s_2) = 0$$

$$\left.\begin{array}{c} s_1 \\ \\ s_2 \end{array}\right\} = \frac{1}{2}\left\{-(1/T_1 + 1/T_2) \pm [(1/T_1 + 1/T_2)^2 - 4(1 + K_1K_2)/T_1T_2]^{1/2}\right\}$$

s_1, s_2 are real if $(1/T_1 + 1/T_2)^2 > 4(1 + K_1K_2)/T_1T_2$

s_1, s_2 are equal if $(1/T_1 + 1/T_2)^2 = 4(1 + K_1K_2)/T_1T_2$

s_1, s_2 are complex if $(1/T_1 + 1/T_2)^2 < 4(1 + K_1K_2)/T_1T_2$

(c)

Fig. 5-8. (a) Two first-order noninteracting processes are represented by their real poles in the complex plane. (b) The same system with feedback added. (c) The characteristic equation for the system in (b) may have either real or complex roots depending upon the relative values of the system parameters. (d) With feedback and some moderate value of gain, the poles move from their original location ($-1/T_1$ and $-1/T_2$) to s_{11} and s_{21}. With further increase in gain the poles may become complex, as at s_{12} and s_{22}.

a. We continue our examination of the dynamic behavior by considering the effect of loop gain upon the root (pole) location. The two first-order processes in (a) are not provided with feedback, and because the processes are noninteracting, the behavior of X_3 is given by the sum of the two exponential modes of the individual processes. The poles are on the negative real axis.

b. With the addition of feedback, as in (b), the characteristic equation is no longer $(T_1s + 1)(T_2s + 1) = 0$, and the roots as obtained from the quadratic formula are quite different from $-1/T_1$ and $-1/T_2$. The loop gain K_1K_2 is seen to have a major effect upon the pole location. Without feedback the poles are $-1/T_1$ and $-1/T_2$, but with feedback and K_1K_2 at some moderate value, the poles may move to s_{11} and s_{21}, for example. These two poles are still real, and the modes are therefore exponential but with changed time constants. Furthermore, the two poles remain symmetrically placed with respect to their average value, that is, $-(1/T_1 + 1/T_2)/2$, with the result that the feedback has caused one of the poles to increase and the other to decrease in magnitude.

c. Examination of the quadratic formula will show that the roots remain real for all values of K_1K_2 that satisfy the inequality $4(1 + K_1K_2)/T_1T_2 < (1/T_1 + 1/T_2)^2$. However, when K_1K_2 becomes large enough to reverse this inequality, that is, to make $4(1 + K_1K_2)/T_1T_2 > (1/T_1 + 1/T_2)^2$, then the quantity under the radical sign becomes negative and the roots are written as

$$s_{12}, s_{22} = -(1/T_1 + 1/T_2)/2$$
$$\pm \left[\sqrt{-1} \left(4(1 + K_1K_2)/T_1T_2 - (1/T_1 + 1/T_2)^2 \right)^{1/2} \right] \Big/ 2,$$

or as $s_{12}, s_{22} = \alpha \pm j\omega$. $j\omega$ is an imaginary number and j is equal to $\sqrt{-1}$.

d. One thus recognizes two distinct cases regarding the roots (or modes) of a second-order system. Either the two roots are real (and in this instance negative) and both modes are exponential, or the roots are complex, in which case the mode is a damped sinusoid as described in the next section. The intermediate case (two equal real roots) is simply the dividing line between these two cases and is of no great interest in itself.

e. All of the roots described up to this point have had negative real parts, and have therefore represented decaying transient terms. Roots with positive real parts are also possible, and will be discussed in a later chapter.

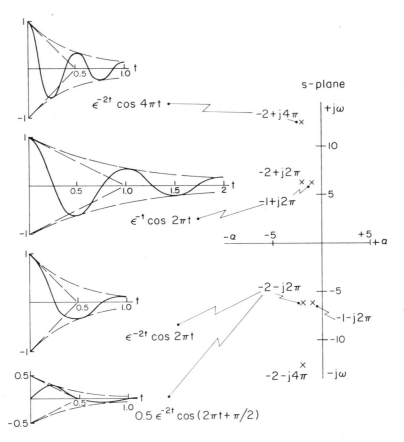

Fig. 5-9. Damped oscillatory modes and their associated poles in the complex s-plane. The complex poles always appear in conjugate pairs (such as $-1 \pm j2\pi$), and each pair gives rise to one such oscillatory mode. The bottom two curves both depict the same mode $(-2 \pm j2\pi)$, but differ in magnitude and phase angle because of a difference in the initial conditions.

5 FEEDBACK SYSTEMS; DYNAMIC BEHAVIOR

a. This figure shows three pairs of conjugate complex roots on the s-plane, and the corresponding damped oscillatory modes. The two roots of a conjugate pair have equal real parts, and imaginary parts that are equal in magnitude but of opposite sign. The two roots of the pair are thus symmetrically located with respect to the real axis, and may be written as s_{12}, $s_{22} = \alpha \pm j\omega$.

b. A pair of conjugate complex roots, $\alpha \pm j\omega$, results in a damped sinusoidal transient having an angular velocity of ω rad/sec and a damping envelop which is an exponential curve having the time constant $1/\alpha$. Thus, complex roots located far to the left of the vertical axis have a large negative α and a small time constant so that the oscillation would rapidly disappear. Furthermore, complex roots having large imaginary components would result in oscillations of high frequency (Trimmer, 1950).

c. The complex oscillatory mode has an angular velocity and a damping time constant established by the coordinates of the root. However, when that mode appears in the dynamic response to a specific disturbance it will take the form $C \exp(\alpha t) (\cos(\omega t + \psi))$, where C sets the magnitude of the oscillations and ψ their phase. The constants C and ψ are calculated for a specific disturbance and set of initial conditions, but whatever their values may be, they do not change the character of the mode. Thus the lower two transients are obtained from the same pair of complex roots, but they differ from each other in phase and magnitude, as might be the case with different disturbances and initial conditions.

d. All the transients in this figure happen to be expressed in terms of the cosine function. This is simply a matter of convenience, inasmuch as $\cos \omega t = \sin (\omega t + \pi/2)$, and any oscillatory mode may be described by either a sine or a cosine function, together with an appropriate phase angle.

e. In a previous discussion of the damped oscillatory response (Section 3.10) this mode of behavior was ascribed to the flow of energy between two forms, kinetic and potential. However, as the system becomes more complex, and especially when feedback is added, this simple explanation is no longer adequate. Feedback results in all variables being coupled with each other, and all variables will exhibit in varying degrees the same oscillatory mode. The energy flow that is associated with the oscillation is more generalized, and may well include participation by many components in the system.

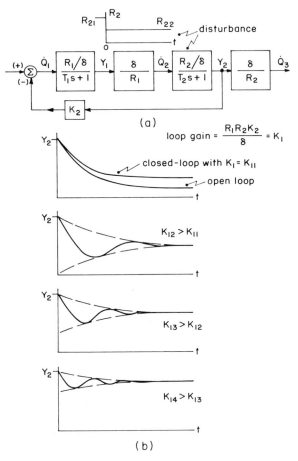

Fig. 5-10. Damped oscillatory response of a second-order regulating system. (a) Block diagram showing the disturbance as a decrease in R_2. (b) Behavior of the quantity Y_2 for increasing values of the loop gain $R_1 R_2 K_2 / \delta = K_1$.

a. In the discussion of the first-order systems in Figs. 5-1 and 5-2, it was shown that the addition of feedback and an increase in the loop gain both had the effect of increasing the speed of response following a disturbance. This was described as a reduction in the system time constant. In systems having two or more compartments, the consequences are similar, but now oscillatory modes may appear even though no oscillations were possible in the open loop.

b. This figure represents a system essentially the same as those in Figs. 5-5 and 5-6, although the block diagram has been drawn somewhat differently. The responses follow a step function decrease in R_2 for progressively larger values of loop gain K_1. For the value K_{11}, the response consists of two exponential modes, and for all larger values of K_1 the response contains a damped sinusoid. Note that the steady-state perturbation in Y_2 becomes less with the increase in K_1; this is a consequence of the decrease in system sensivitity.

c. The initial rate of change of Y_2 is the same for all cases because it depends only upon the initial value of Y_2, the system parameter A_2, and the change in R_2. At the first instant, dY_2/dt does not depend on the presence of feedback nor upon the loop gain.

d. In comparing the oscillations that appear as the loop gain is increased, it will be found that their frequency increases, or the period becomes shorter. However, the time for decay remains the same, the envelopes all have the same exponential character, and the same time constants. Thus, with an increase in the gain there is less damping per cycle, but no change in the damping per unit of time.

e. The oscillations, here shown for the variable Y_2, also appear in all the other variables, and with the same angular velocity and damping time constant. Furthermore, although these oscillations were induced by a change in R_2, any other disturbance would have similar consequences. The existence and character of the oscillations depends solely upon the relative values of the several system parameters following the disturbance.

f. The increased gain postulated here might be obtained by shifting the fulcrum to the left. With the increased gain, the valve movement becomes greater for a given change in Y_2. Thus, there is a tendency to overcorrect and overshoot the final steady-state values. This viewpoint is sometimes used to explain the oscillatory behavior.

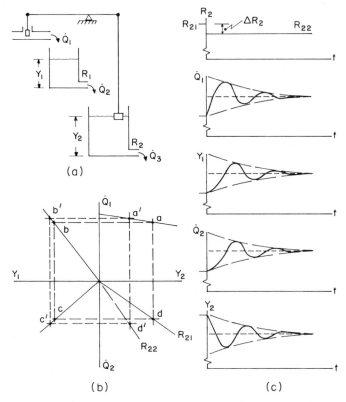

Fig. 5-11. Signal transmission through a second-order feedback system following a step function decrease in R_2. (a) Block diagram of the system. (b) Steady-state characteristics showing the initial operating point (a, b, c, d) and the final operating point (a', b', c', d'). (c) Responses of the dependent variables, all of which have the same frequency and damping, but differ in phase.

Signal Transmission through a 5.12
Feedback System Having an
Oscillatory Response: Fig. 5-11

a. This section is devoted to a further examination of the oscillatory case to show how the oscillations of the several variables are related to each other. The discussion is based upon two simple propositions already developed in earlier sections. (1) The component processes are connected so that the output of one serves as the input to the next, and this relation holds right around the feedback loop. (2) It takes time for a signal to propagate through a process, so the output signal will appear delayed in phase compared to the input.

b. The system shown in this figure is exactly the same as that in Figs. 5-5 and 5-6. The disturbance considered is a step function reduction in R_2.

c. Part (b) of the figure shows the effect of this change on the steady-state values of the dependent variables. Although Y_2 decreases, all the other variables increase as a consequence of the change. The transient behavior is shown in (c), but note that for clarity these curves have been drawn to different scales than those used in (b).

d. The immediate effect of reducing R_2 is a discontinuous increase in $\dot{Q}_3$ (not shown in this figure) and as a consequence, Y_2 starts to fall. The resulting change in $\dot{Q}_1$ is strictly proportional to that of Y_2, but of the opposite sign. This is equivalent to stating that the oscillation in $\dot{Q}_1$ is $180°$ out of phase with that of Y_2.

e. The oscillation in Y_1 (as well as in $\dot{Q}_2$) lags $\dot{Q}_1$ by approximately $90°$. This phase difference, or time interval, arises from the dynamic properties of the first compartment, that is, from the time required to empty or fill the tank.

f. Similar observations may be made for the second compartment, where Y_2 lags behind $\dot{Q}_2$ by approximately $90°$. The successive phase shifts that appear as the signal is propagated through the two compartments may also be considered in a somewhat different light. All the system variables oscillate at the same frequency, and this frequency is "set" by the system to produce the phase shifts described above. The total phase shift is only approximately $360°$ for this system. The consequences of a greater phase shift are developed in Chapter 7.

g. The oscillatory transient must be looked upon as a property of the entire system, and not of just a single parameter or component. The mode (or modes) of oscillation depend upon the relative values of all the system parameters, and are the same for all system variables, as has been stated previously.

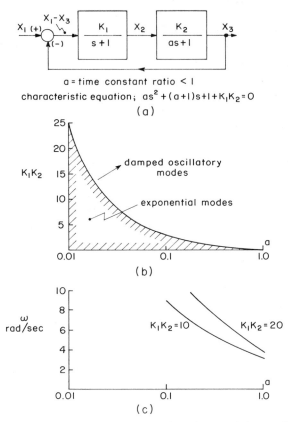

a = time constant ratio < 1

characteristic equation; $as^2 + (a+1)s + 1 + K_1K_2 = 0$

(a)

(b)

(c)

Fig. 5-12. (a) Second-order feedback system and its characteristic equation. (b) Values of loop gain lying above the shaded region produce damped oscillatory modes. For a given time constant ratio a, the value of K_1K_2 must lie in the shaded region to produce exponential modes. (c) For a given value of loop gain, the angular velocity ω of the oscillatory mode decreases as a is increased.

a. The compartment time constants and their influence upon the transient modes of a feedback system will be considered in this section. In particular, we wish to show the manner in which the time constant *ratio* affects the dynamic behavior.

b. As an example, part (a) of the figure shows two first-order processes connected to form a second-order feedback system. For convenience, the largest time constant has been given the value unity, and the smallest the value $a < 1$, so that the ratio is a. The characteristic equation shown in the figure is readily developed from the transfer functions, and the roots are obtained from application of the quadratic formula.

c. We are here concerned with the magnitude of the loop gain, K_1K_2, at the dividing point between real and complex roots. The results are shown in (b), where this value of K_1K_2 is plotted as a function of a. Combinations of K_1K_2 and a lying above the line yield complex roots and a damped oscillatory mode.

d. From the figure it is clear that as the time constants become more nearly equal, the loop gain required to maintain exponential modes becomes very small. On the other hand, as one time constant becomes very small with respect to the other, the gain required to produce an oscillatory mode may become quite high.

e. We have previously seen that for a given time constant ratio, the angular velocity associated with the transient mode increases as the loop gain is increased. Another view of these relations is afforded by part (c), where the effect of a upon the angular velocity ω is shown for a constant value of K_1K_2. The magnitude of ω decreases as a approaches unity.

f. The time constants mentioned in the above discussion are those of the individual compartments, previously termed the "open-loop" time constants, the effect of their ratio upon the dynamic response of the closed loop (the feedback system) has been the major concern.

g. The relations described here for a second-order system may be extended to systems of higher order. The more nearly equal the several (open-loop) time constants are, the smaller the loop gain required to produce oscillatory modes. Under such conditions, there is a better opportunity for the energy flow between compartments to oscillate. The relations are not at all affected by the location, or the sequence, with which the several compartments appear in the system, but only upon the time constant ratios.

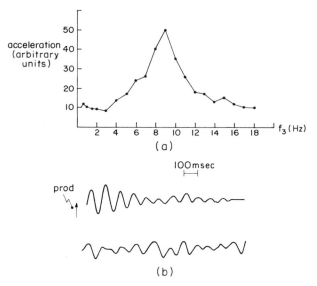

(a)

(b)

Fig. 5-13. Studies in finger tremor. (a) Frequency spectrum of the ongoing tremor in a finger. From "Physiological Tremor" by O. C. J. Lippold, Copyright © March, 1971 by Scientific American, Inc., all rights reserved. Note that this is a plot of measured acceleration. (b) Upper curve shows the finger displacement following a pulse stimulus (prod). Lower curve shows the ongoing tremor without a prod stimulus. Both curves are to the same scale. From Lippold (1971) by permission of the author. (Note: this figure and others throughout the text which depict experimental observations have been redrawn to emphasize those aspects pertinent to the exposition that accompanies them. The reader is urged to consult the original papers for further information.)

a. Although a muscle by itself exhibits an overdamped response with no oscillatory behavior, when observed as a component in the monosynaptic spinal reflex (Section 4.20), the response is underdamped and highly oscillatory. Evidence attributing this oscillatory behavior to the feedback is presented below (Lippold, 1970).

b. Physiological tremor is associated with voluntary control of all skeletal muscles. In the finger, this tremor is found to have a frequency spectrum centered around 9 Hz. The fact that the spectrum for acceleration peaks as shown in (a) implies that the force also peaks, and this spectrum represents muscle force.

c. Observation of a single response following a prod stimulus as in (b) shows a well-defined damped sinusoid. Averaging a number of these responses gives a curve of the same character and enables one to follow the damped sinusoid for more cycles.

d. The lower curve in (b) shows the ongoing tremor (without stimulus), which is seen to contain many cycles of approximately the same frequency as observed in the upper trace. Although the amplitude of this tremor is very small, it has been shown to be adequate for operation of the muscle spindle.

e. The hypothesis under examination is that the monosynaptic spinal reflex is a lightly damped system, and that the ongoing tremor represents the responses to a succession of stimuli appearing at random. This is supported by the fact that in animals, sectioning the afferent pathway at the dorsal roots abolishes the tremor. In humans, tabes sometimes abolishes transmission in the afferent nerves, and such patients usually show no sign of these oscillations.

f. Cooling and warming the muscle is found to change the oscillatory frequency, and this change is essentially the same for both the ongoing tremor and the oscillations following a mechanical disturbance.

g. Simultaneous recording of muscle action potentials and finger tremor show the two signals to be synchronized, thus establishing a relation between active neuromuscular events and the oscillations.

h. Physiological tremor is readily distinguished from other oscillations by the characteristic frequency. The mechanical resonant frequency of the finger is of the order of 30 Hz, and Parkinsonianism frequency is about 5 Hz, while the cardiac frequency is around 1 Hz.

i. The above, as well as other evidence, serves to support the notion that tremor is simply the oscillation in a lightly damped feedback system. If the ongoing tremor in (b) is averaged over a number of sweeps, the oscillations tend to disappear.

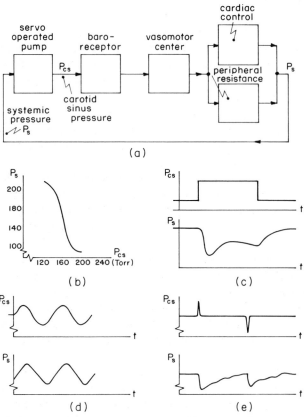

Fig. 5-14. Properties of the baroreceptor reflex for regulating blood pressure. (a) Simplified block diagram showing the experimental setup in which a servo operated pump is inserted into the carotid artery so as to effectively open the loop and provide for the introduction of a test signal to P_{cs}. (b) Steady-state characteristic. (c) Responses to positive and negative step functions. (d) Sinusoidal response. (e) Responses to positive and negative pulses.

a. The measurement of dynamic behavior of the components in a feedback system is very difficult if the "loop" cannot be opened. So far, this has proven to be difficult in physiological systems, owing largely to the multiplicity of loops, as well as the necessity of terminating the cut ends in a physiological manner. The baroreceptor reflex (Section 4.19) is one example where this has been done rather successfully (Scher and Young, 1963).

b. The greatly simplified diagram (a) is based upon Fig. 4-18, with the addition of a servo operated pump inserted into the carotid sinus artery. This serves to isolate the arterial (systemic) pressure P_s from the carotid sinus pressure P_{cs} and permit the latter to be independently controlled. The sinus nerves were kept intact, but signals from aortic receptors were blocked by denervation. The assumption made was that this destroyed all neural pathways affecting arterial pressure except those from the carotid sinus, thus making the experiments truly "open loop."

c. The servo operated pump forced the input pressure P_{cs} to follow an arbitrary signal. P_s was measured to give the output signal, thus obtaining the complete transmission through the loop.

d. Measurement of steady-state gain yielded curves such as (b), having the negative slope expected for a regulatory system. The maximum gain showed a wide variation between individual animals, but an average of around five in cats and near two in dogs.

e. The response to step changes in P_{cs}, while somewhat variable, tended to be of the form shown in (c). The response to a rise in pressure usually showed considerable overshoot, whereas the response was monotonic following a pressure fall.

f. The sinusoidal behavior as in (d) showed relatively little distortion, and of course a 180° phase lag at very low frequency. With an increase in frequency, the angle increased to 360° and somewhat beyond. The amplitude portion of the spectrum exhibited a small resonant peak in the vicinity of 0.03 to 0.05 Hz (Levison, Barnett, and Jackson, 1966).

g. The response to a pulse input in P_{cs} exhibited a feature not encountered with linear systems, and yet quite commonly observed in physiological processes. Both positive and a negative pressure pulse resulted in decreases in the P_s that were not greatly different from one another (e). This phenomena, related to the asymmetrical character of many response patterns, has been termed unidirectional rate sensitivity and has been examined in a number of physiological systems (Clynes, 1969).

5.15   BARORECEPTOR REFLEX   157

As already mentioned in Section 5.3, it is impossible to identify the existence of feedback from observations made at the terminals of a system, that is, of specific input and output variables. This fact has considerable bearing on the general problem of identification, and since this is a central question in homeostatic processes, the matter requires further elaboration. Furthermore, it has a direct relation to the interpretation of experimental observations on intact systems.

In the first place, reference to the analysis of linear dynamic behavior as outlined in Section 3-16 will show that no reference is made regarding the existence or absence of feedback. The final equation of motion is an ordinary differential equation with constant coefficients, and an equation of exactly the same form would be obtained whether or not the system contained feedback pathways. However, the initial set of equations, those written directly from the physical processes, do give evidence of feedback if such exists, and this may be evaluated by application of the rules given in Section 4.3. The analysis procedure, starting with this set of equations, reduces them to a single differential equation (a different one for each variable) and in the process of combining the several equations, all evidence for the existence of feedback disappears. This may be regarded as the penalty paid for the mathematical abstraction necessary to predict the system behavior to some future disturbance. Despite this loss of information, the properties of the system as modified by feedback are still retained within the single equation of motion, but it is no longer possible to infer that these properties are derived from feedback as the identification process requires.

The situation described above may be stated in somewhat different terms by observing that a given differential equation might describe equally well a system without feedback as one having feedback. This is equivalent to saying that in principle, for every feedback system there is an equivalent nonfeedback system with identical properties. This does not mean that feedback has no effect, but rather that feedback makes a given set of components appear as if they were quite different. If it were possible to synthesize a system from an unlimited store of all types of linear components, then in principle, one could design a system having no feedback, but having the properties of one that did. Even in the area of engineering design, such a freedom of choice is practically never available, and of course in the domain of living systems the hypothesis is essentially devoid of meaning. Nevertheless, this is an essential viewpoint in the formulation of a model, particularly if that model is to be at all isomorphic with the physiological processes.

There have been a number of instances in the literature in which systems having no feedback, as defined by the rules of Section 4.3, were treated as though they did. The error in this procedure is that by following the same reasoning, one comes to the conclusion that all physical systems may be viewed as feedback systems, and there is nothing to distinguish the important class of physical systems in which feedback is physically present from those lacking this connection. The above argument necessitates the observation that feedback must be considered a physical concept, and not a mathematical one. This is the basic reason why the earlier definition of feedback in Chapter 4 was couched in operational or physical terms, so that the identification of feedback requires examination of the physical system.

In the event that additional information regarding the system is available, then the identification process from external measurements can be carried further. For instance, if it is known that all components within a system are first-order processes having simple exponential modes, then a system without feedback but containing any number of such compartments can only have a response consisting of the sum of exponential modes. However, if experiment shows that oscillatory modes are present, this can only be the consequence of feedback loops within the system that are able to generate oscillatory modes after the fashion of Fig. 5-10.

A second instance is that of a system which exhibits instability, as will be discussed in Chapter 7. It is shown that instability exhibited by a system of stable components can only arise in the presence of feedback. It is thus apparent that with some additional information regarding the system components, it is possible to infer the presence of a feedback pathway from observations made at the terminals.

Aside from the desire to construct an isomorphic model, there are other reasons for needing to develop a correct system representation. Chief among these is the problem of interpretation of experimental observations. Consider a system in which one variable is stimulated and the response of another is observed. If no feedback is present, the stimulus–response pair correctly describes the component or components across which the experiment is performed. However, with feedback present, the stimulus may not be the total input to the first component since another signal appears at this point as the result of feedback. Hence without additional precautions and information regarding the system, the measured response may be misinterpreted.

1. In addition to decreasing the sensitivity, negative feedback significantly modifies the dynamic properties of a system. In general, the effect is to speed up the response, introduce oscillatory responses, and decrease the damping of the transient components.

2. Negative feedback applied to a first-order process reduces the response time constant, as well as the closed-loop gain, compared to the values they would have in the absence of feedback. These reductions are by the same factor, $1/(1 + K)$.

3. The closed-loop transfer function, which exhibits the above reductions, may be considered as an equivalent open-loop system. The basic character of the transfer function remains unchanged; it is thus impossible to tell from measurements made at input and output terminals only, whether or not the system does in fact contain feedback.

4. A change in the operating point of a first-order regulator results in single-exponential transients in all the dependent variables. These transients have identical time constants, but in general are of different amplitude and sign.

5. The effects of a disturbance to a first-order system may be followed on the steady-state diagram, from which the initial and final value of all variables can be found. Points on this diagram represent the state of the system, and the path followed by this representative point is termed a trajectory.

6. In a second-order feedback system, all dependent variables respond to any disturbance with the same transient modes. If these are exponential modes, they will usually appear with different amplitudes and sign in the several variables so that with only a cursory inspection of the responses, they may appear to be quite unlike.

7. Disturbances are quite likely to alter the magnitude of the system parameters, but the same basic relations described above still hold. However, the response to such a disturbance must be calculated with parameters of the disturbed system.

8. The exponential mode of response obtained with a first-order process is associated with a real root of the characteristic equation. The modes may thus be represented by points on the s-plane, the exponential modes being points on the negative real axis.

9. Two first-order processes forming a negative feedback system have a characteristic equation of the second degree, the roots of which may be either real or complex. With low gain, the roots are real, but approach each other as the loop gain is increased.

10. Complex roots of the characteristic equation always appear as conjugate pairs, and each pair gives rise to a damped oscillatory mode of behavior. The abscissa of the roots is the reciprocal of the damping time constant, and the ordinate is the angular velocity.

11. With sufficient gain, every second-order feedback system may be made to exhibit damped oscillations following any disturbance. In this oscillatory response mode, a further increase in gain increases the oscillatory frequency, but the exponential damping envelop remains the same.

12. When transient oscillations appear in any system variable, they appear in all the dependent variables with the same frequency and exponential decay. However, a comparison of the several variables shows that they do vary in amplitude and in relative phase.

13. The dynamic behavior of a feedback system is dependent upon the ratio(s) of the open-loop time constants in the following sense. The closer these ratios are to unity, the smaller the gain required to produce oscillatory modes. Although an increase in any of the open-loop time constants will produce a feedback system with slower response, the oscillatory character of the response depends upon the time constant ratios.

14. Analysis of the finger tremor supplies strong evidence for the notion that the stretch reflex is a feedback system with relatively little damping. This conclusion is supported by the observation that the finger position exhibits extended oscillations following a pulse stimulus.

15. The behavior of the blood pressure regulator termed the baroreceptor reflex has been studied by opening the loop at the carotid artery. The loop gain for the cat is of the order of five; this maximum occurs at an arterial pressure of around 170 Torr. Although the system seems to be reasonably linear, as judged from the sinusoidal response, for other disturbances there are some surprising nonlinearities.

16. The identification of feedback requires a knowledge of the system components and their interconnections; it cannot be obtained from external measurements on the intact system, because the behavior of the system is not qualitatively different without and with feedback. Furthermore, input–output measurements on any component in a feedback system rarely provide information that will correctly characterize that component because of the feedback signal that appears along with the stimulus at the input.

5.17   EPITOME   161

1. For the system of Fig. 5-1:
(a)   Sketch the behavior of $\dot{Q}_i$ and $\dot{Q}_2$ for the disturbance shown.
(b)   Sketch the behavior of Y_1 to a step function increase in P_1 (that is, from P_{11} to P_{12}) in a manner similar to that shown.
(c)   Develop the open- and closed-loop transfer functions for the system in this figure. Show that $T_{12} = T_{11}/(1 + K)$ as suggested in the figure.
(d)   Write the differential equation and the characteristic equation that describe this system. Use the variable Y_1.

2. For the system in Fig. 5-2:
(a)   Verify the closed-loop transfer function shown.
(b)   What is X_2 in terms of the relations shown in Fig. 5-1?

3.   (a)   What is the physical significance of the intercepts Y_{15} and Y_{16} in Fig. 5-3b?
(b)   Sketch the time response for the system in Fig. 5-3 following a sudden *increase* of L_1 from L_{12} to L_{11}. Note that $\dot{Q}_i$ may drop to zero, but physically cannot become negative.
(c)   Verify that the loop gain is approximately 9.

4.   (a)   Construct a steady-state diagram (after the manner of Fig. 4-2) for the system shown in Fig. 5-5. Show the operating point both before and after the disturbance.
(b)   Compare the dynamic behavior of the dependent variables in Figs. 5-5 and 5-6. Account for the differences on physical grounds.

5.   Sketch the dynamic behavior of $\dot{Q}_2$ and $\dot{Q}_3$ that corresponds with the transient shown in Fig. 5-6.

6.   (a)   Assume that $(1/T_1 + 1/T_2)^2 < 4(1 + K_1K_2)/T_1T_2$ in Fig. 5-8. Show that the roots s_1, s_2 are

$$\frac{-(1/T_1 + 1/T_2)}{2} \pm j\left(\frac{(1 + K_1K_2)}{T_1T_2} - \frac{(1/T_1 + 1/T_2)^2}{4}\right)^{1/2}$$

where $j = \sqrt{-1}$.
(b)   Show that for any values of T_1 and T_2, K_1K_2 can always be made large enough to yield a damped oscillatory solution.
(c)   Show that if the ratio T_1/T_2 is *increased,* the value of K_1K_2 necessary to produce complex roots is also increased.

7.   Sketch, in the manner used in Figs. 5-7, 5-8, and 5-9, the modes associated with the following roots:
  (a) $s_1 = -5$;
  (b) $s_2 = +3$;
  (c) $s_3, s_4 = -4 \pm j4\pi$;
  (d) $s_5, s_6 = +2 \pm j4\pi$.

8.   Sketch the behavior of the system variables in Fig. 5-11 following a step function decrease in R_1, that is, $-\Delta R_1$.

9.   (a) Derive the equation that separates damped oscillatory from exponential modes for the curve in Fig. 5-12b between K_1K_2 and a.

  (b)   Derive the equation for Fig. 5-12c and confirm the coordinates of one point on each curve.

10.   The statement was made in Section 11c that the initial value of dY_2/dt following the disturbance is independent of gain K_1. Justify this statement on physical grounds. Calculate the value of this derivative at $t = 0+$ in terms of the other system parameters. (Hint: Write the differential equations for the system in Fig. 5-10.)

11.   In the single-compartment process of Fig. 3-4, consider $\dot{Q}_1$ as the input and $\dot{Q}_0$ as the output variable. Write the transfer function for the process, $\dot{Q}_0/(\dot{Q}_1 - \dot{Q}_0)$ in terms of the integral $1/s$. Show that this process could be represented by the feedback system of Fig. P5-11.

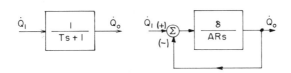

Fig. P5-11

References

Clynes, M. (ed.) (1969). Rein control, or unidirectional rate sensitivity: a fundamental dynamic and organizing function in biology. *Ann. N. Y. Acad. Sci.* **156,** Art. 2, 627–968.
Eckman, D. P. (1966). "Automatic Process Control." Wiley, New York.
Johnson, E. F. (1967). "Automatic Process Control." McGraw-Hill, New York.
Levison, W. H., Barnett, G. O., and Jackson, W. D. (1966). Nonlinear analysis of the baroreceptor reflex system. *Circul. Res.* **18,** 673–682.

Lippold, O. C. J. (1970). Oscillation in the stretch reflex arc and the origin of the rhythmical, 8–12 Hz component of physiological tremor. *J. Physiol.* **206**, 359–382.

Lippold, O. (1971). Physiological tremor. *Sci. Amer.* **224**, Mar. 65–73.

Machin, K. E. (1964). Feedback theory and its application to biological systems. Symp. Soc. *Exper. Biol.* **28**, 421–445.

Milsum, J. H. (1966). "Biological Control Systems Analysis." McGraw-Hill, New York.

Scher, A. M., and Young, A. C. (1963). Servoanalysis of carotid sinus reflex effects on peripheral resistance. *Circul. Res.* **12**, 152–162.

Trimmer, J. D. (1950). "Response of Physical Systems." Wiley, New York.

It will have become apparent that with increasing system complexity, the explanation of behavior becomes more difficult. This complexity arises from the presence of a number of component processes, and the fact that an increased number of dependent variables is required to describe the system behavior. The many dependent variables are related to each other in ways that are far from obvious. Furthermore, when the system includes feedback paths, so that it becomes impossible to speak of simple cause and effect relationships, it also becomes increasingly difficult to predict, or even interpret, system behavior without additional conceptual tools.

The use of a sine wave (*a sinusoid*) as a disturbance to a complex system offers certain advantages in the study of system behavior. Many physical systems, when disturbed by a sinusoid, will respond in a sinusoidal manner so that the several dependent variables will exhibit continuing sinusoidal oscillations of the same frequency. However, the relations between the oscillations of the several dependent variables provide a convenient means of assessing system behavior.

The sinusoidal disturbance may be introduced in an experimental situation and the oscillatory responses actually measured. This is a widely used technological tool and is also being introduced in biological studies, as will be discussed later. On the other hand, whether or not it is used experimentally, the method of sinusoidal analysis is a powerful conceptual tool, and as such provides a different insight into system behavior (Pringle and Wilson, 1952).

The sinusoidal method of analysis is not directly applicable to all systems, and thus is not a panacea. However, it is valid for many physical processes, and for those systems that are not completely amenable to sinusoidal analysis, it still provides an insight not obtainable in other ways (Stoll, 1969).

Although many biological processes do in fact oscillate continually, this discussion is not directed solely to that class of systems. The immediate application of sinusoidal methods will be to systems which either show no inherent oscillatory behavior, or if they do exhibit transient oscillations, the transient is rapidly damped out (Stark and Sherman, 1957).

The sinusoid is the simplest form of a *periodic function,* which is a quantity that undergoes a cyclic change with time, and repeats that cyclic variation indefinitely. The sinusoid occupies a distinctive place in the study of periodic functions, that is, any periodic function can be described as the sum of a number of sinusoids. Thus the sinusoid is a fundamental building block in the description of periodic functions of arbitrary shape.

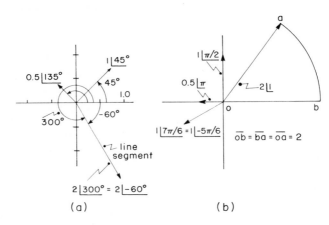

Fig. 6-1. Directed line segment with its angular measure in (a) degrees, and (b) radians.

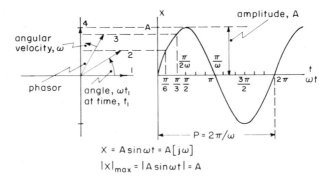

$$X = A\sin\omega t = A[\,j\omega\,]$$
$$|X|_{max} = |A\sin\omega t| = A$$

Fig. 6-2. A sinusoid, $X = A$ sin ωt, and its phasor representation. The fact that x is a sinusoidal quantity is shown formally by the functional notation, $A[\omega t]$. The period P is the time required for one cycle, and the frequency in hertz can be expressed by either of the following expressions, $f = 1/P = \omega/2\pi$.

a. A geometric interpretation of a sinusoid may be obtained with the aid of a *directed line segment* emanating from the origin of a coordinate system. Each line segment is defined by its length and the angle it makes with the horizontal axis. The line segment $A\angle\theta$ has a length of A units and makes an angle θ with the positive horizontal axis.

b. The angles could be measured in either a counterclockwise or clockwise sense, but it is common usage to designate the counterclockwise sense as positive. Note, however, that the line segment $2\angle300°$ may also be described as $2\angle-60°$.

c. Angles may be measured in degrees or in radians. One radian is defined as that angle which subtends a circular arc equal to the radius. One complete revolution (360°) is then 2π radians, and 90° is $\pi/2$ radians.

d. Consider a line segment of length A, *rotating* at a constant angular velocity of ω radians per second ($A\angle\omega t$). If ω is positive, the line rotates counterclockwise, and if $\omega = 2\pi$ radians per second (rad/sec), the line makes one complete revolution in one second. A line segment rotating at a constant angular velocity is termed a *phasor*.

e. Inasmuch as a phasor is rotating, it can only be shown as occupying certain positions instantaneously at prescribed angles, or instants of time. In Fig. 6-2, the phasor successively occupies positions 1, 2, 3, 4, etc., corresponding to angular positions 0, $\pi/6$, $\pi/3$, and $\pi/2$ radians at successive instants of time. If $\omega = 2\pi$ rad/sec, then these positions correspond to 0, $\frac{1}{12}$, $\frac{1}{6}$, $\frac{1}{4}$ sec respectively. The angle describing the phasor at a given instant is termed the *phase angle* or *phase*.

f. A sine wave may be constructed from a phasor by taking the vertical projections at successive instants of time. The sinusoid $A\sin\omega t$ has the magnitude 0 at $\omega t = 0$, rises to A at $\omega t = \pi/2$, and falls to zero again when $\omega t = \pi$. The *amplitude* of a sinusoid is its maximum magnitude, that is, $|X|_{max} = A$.

g. The horizontal scale for the sinusoid may be expressed as an angle (ωt) or, since ω is a constant, in units of time. The time required for one complete revolution (one cycle) is termed the *period* (designated P, in seconds). The period P in angular units is always 2π radians.

h. The rotating line segment, or phasor, provides a geometric and conceptual aid in visualizing a sinusoid and the several quantities used to describe it. However, one should not expect to find a physical counterpart for the phasor in a physical or biological problem.

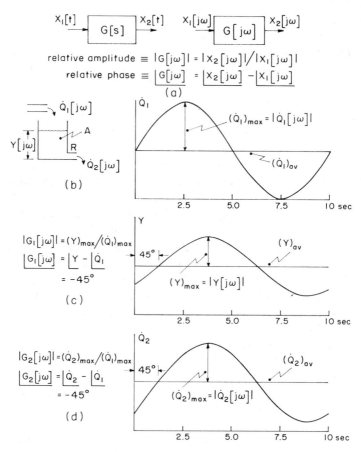

Fig. 6-3. Sinusoidal transmission through a single compartment at a frequency (f) of 0.1 Hz, or an angular velocity (ω) of 0.628 rad/sec. The amplitude and phase of each dependent variable (Y, $\dot{Q}_2$) are functions of ω. The size of the compartment shown in this figure has been arbitrarily selected to be such as to yield a phase angle of $-45°$ at 0.1 Hz.

a. In earlier discussions, a given process has been described by its transfer function, written as $G[s]$. With suitable mathematical techniques (Laplace transform) this transfer function enables one to calculate the response, $X_2[t]$, for any imposed disturbance $X_1[t]$. This may be described as the transformation of one time-dependent variable into another. This chapter is concerned with the special case in which both X_1 and X_2 are sinusoidal quantities, for which one writes $X_1[j\omega]$, $X_2[j\omega]$, and the transfer function becomes $G[j\omega]$. As this expression implies, $G[j\omega]$ is obtained directly from $G[s]$ by substituting $j\omega$ for s at every point in the transfer function (Trimmer, 1950).

b. The transmission of a sinusoidal signal through a process is described in terms of the relative amplitude and phase of the input and output signals as given in (a). The amplitude of a sine wave may be written as $(X_1)_{\max}$, or as $|X_1[j\omega]|$. Note that $X_1[j\omega]$ denotes a sinusoidal variable, whereas $G[j\omega]$ denotes a (transfer) function for sinusoidal quantities. A lagging phase angle has a negative sign as in (c) and (d).

c. In the discussion of sinusoidal transmission, it is assumed that any transient behavior that might appear at the outset has disappeared, so that the process is in a sinusoidal steady state with all dependent variables undergoing sinusoidal oscillations of the same frequency. Furthermore, we are not concerned here with the mechanism by which sinusoidal oscillations are generated.

d. In this first-order flow process, both Y and $\dot{Q}_2$ oscillate in phase with each other, but both have a phase lag with respect to $\dot{Q}_1$. Negative values of Y and $\dot{Q}_2$ have no physical meaning, so that to maintain undistorted sinusoidal variations of all variables, the disturbance has been superimposed upon an average value so chosen that $\dot{Q}_1$ never becomes less than zero. As a general rule, the amplitude of a disturbing sinusoid is kept small with respect to its average value so as to avoid any possible distortion.

e. The phase lag between $\dot{Q}_1$ and Y reflects the fact that time is required to fill and empty the tank. On the other hand, $\dot{Q}_2$ and Y are in phase because they are related by a constant; $\dot{Q}_2 = \delta Y/R$. The phase lag at a given frequency will increase if the size of the compartment is increased.

f. The transmission function may have dimensions, as is the case for $G_1[j\omega]$ in (c), whereas $G_2[j\omega]$ is dimensionless in (d) because both input and output variables are the same kind of physical quantity.

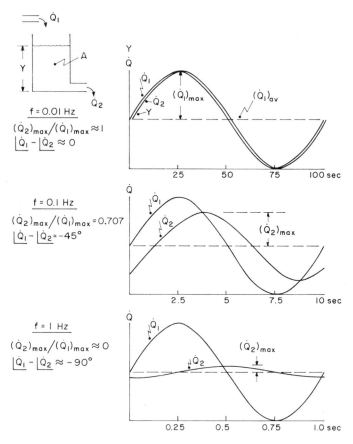

Fig. 6-4. The forced sinusoidal response of one compartment is examined at three different angular velocities, 0.01, 0.1, and 1.0 Hz. Note that the three responses have been drawn with greatly different time scales to clearly show the changes produced at the different frequencies. A more realistic picture would be obtained if all three responses were shown with the same time scale.

6   SINUSOIDAL SIGNALS

a. This section will show how the relative amplitude and phase of the response changes with the frequency of the sinusoidal disturbance. Note that since Y and $\dot{Q}_2$ are related by a constant, the same waveforms could be used for both variables, and the only difference between them is a change in vertical scale.

b. The waveforms for $\dot{Q}_1$ and $\dot{Q}_2$ have been drawn for three different frequencies, selected so as to show the range of frequencies over which the relative amplitudes and phase exhibit the greatest change. At the lowest frequency (0.01 Hz), $\dot{Q}_1$ and $\dot{Q}_2$ are essentially the same in amplitude and phase. For a compartment of any size, a sinusoid can always be chosen having a frequency low enough to yield this same relation. In such a case the time required to change the quantity stored (ΔY) is small, compared to one period of the forced oscillation.

c. In contrast, one finds that at very high frequencies the period is so short that very little change occurs in Y and $\dot{Q}_2$ throughout one cycle. In the figure, at 1 Hz, $\dot{Q}_2$ changes very little during one cycle simply because there is not time available for the movement of any large amount of liquid. The change that does occur is shifted nearly 90° in phase with respect to $\dot{Q}_1$.

d. The changes in relative amplitude and phase that occur as the frequency is increased are termed the *forced oscillatory,* the *sinusoidal,* or the *frequency response* of the process. For all first-order processes, the frequency response is found to have the following properties in relation to the sinusoidal frequency: (a) the amplitude ratio decreases from some constant value at zero frequency (in this instance unity) to zero at infinite frequency, and (b) the relative phase decreases from zero (at zero frequency) toward −90° at very high frequency. The precise manner in which these changes take place, and the corresponding frequency range, serve to uniquely define the process.

e. Some feeling for this relationship may be gained by noting that if the compartment were made twice as large, its time constant would be doubled and changes would occur more slowly. Specifically, the frequencies over which the above transitions would take place are lowered, but in terms of the behavior at very low and very high frequencies, the performance is unchanged.

f. In all of these considerations, it is assumed that all variables undergo strictly sinusoidal variations—that is, the process is linear.

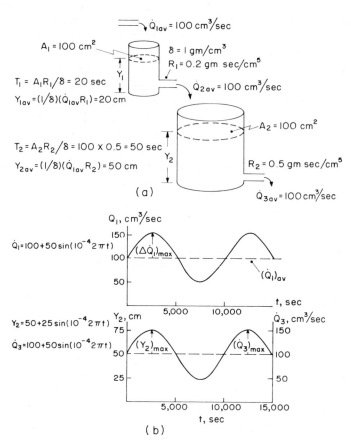

Fig. 6-5. Second-order system having a sinusoidal input with 50% modulation. Inasmuch as the time constants (20 and 50 sec) are very small compared with the period of the sine wave (10,000 sec), the variables Y_2 and $\dot{Q}_3$ have a very small phase lag compared with $\dot{Q}_1$. This angle is assumed negligible and is not shown in the figure.

a. The sinusoidal signal used in the study of physical processes is normally selected to have an amplitude that is small compared with the average value of the variable. The amplitude of the sine wave in this figure has been arbitrarily set at 50% of $\dot{Q}_{1av}$; this is said to constitute 50% *modulation*. The term modulation refers to the manner and extent to which one signal (here the instantaneous value) is caused to vary by multiplying the average value by a second signal (the sinusoid). It is usually desirable to employ small values of modulation (1–5%, for instance) and the value of 50% in this figure was selected solely for clarity of exposition.

b. The sinusoidal frequency (0.0001 Hz) selected is so small that for all practical purposes, the dependent variables are in phase. Although not shown, Y_1 and $\dot{Q}_2$ also execute sinusoidal variations about their respective average values. Note that the period of the sine wave, 10,000 sec, is very large compared to the time constants of 20 and 50 sec.

c. The average values of the several variables have been calculated in the same manner as would be used in the absence of sinusoidal modulation. For this process, the modulation (if less than 100%) has no effect on the average values. Conservation of mass requires that $\dot{Q}_{1av} = \dot{Q}_{2av} = \dot{Q}_{3av}$.

d. In all previous examples, the processes were linear for which the principle of superposition may be applied (p. 47). It is also possible to use the sinusoidal behavior to define a linear system as one that meets all the following requirements: (1) a sinusoidal disturbance produces a sinusoidal response of all variables at the same frequency, (2) a change in the amplitude of the disturbance results in a linearly proportional change in the amplitude of all variables, and (3) the application of two sinusoids of different frequencies produces the same response as would result from the addition of the responses obtained from each sinusoid separately. These requirements mean that there is no interaction between individual sinusoids appearing in the disturbance, and also that no sinusoidal component appears in the response that is not present in the input.

e. A system that fails to meet these requirements is termed *nonlinear*. However, many (but not all) such systems will behave in a linear manner if the disturbing signal is kept small. This is termed a *small-signal analysis* and is the justification for the previous emphasis upon small-percentage modulation.

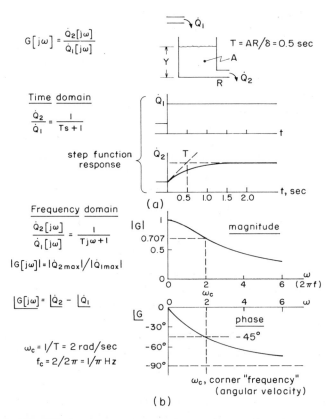

$$G[j\omega] = \frac{\dot{Q}_2[j\omega]}{\dot{Q}_1[j\omega]}$$

$\dot{Q}_1$

$T = AR/8 = 0.5 \text{ sec}$

Y

A

R $\dot{Q}_2$

Time domain

$$\frac{\dot{Q}_2}{\dot{Q}_1} = \frac{1}{Ts+1}$$

step function
response

$\dot{Q}_1$

t

$\dot{Q}_2$ T

0.5 1.0 1.5 2.0   t, sec

(a)

Frequency domain

$$\frac{\dot{Q}_2[j\omega]}{\dot{Q}_1[j\omega]} = \frac{1}{Tj\omega+1}$$

$|G[j\omega]| = |\dot{Q}_{2max}|/|\dot{Q}_{1max}|$

$\underline{/G[j\omega]} = \underline{/\dot{Q}_2} - \underline{/\dot{Q}_1}$

$\omega_c = 1/T = 2 \text{ rad/sec}$

$f_c = 2/2\pi = 1/\pi \text{ Hz}$

|G|

0.707

0.5

0

0   2   4   6   (2πf)
      ω_c

magnitude

ω

0   2   4   6   ω

$\underline{/G}$

-30°

-60°

-90°

phase
- 45°

ω_c, corner "frequency"
(angular velocity)

(b)

Fig. 6-6. Response of a first-order process. (a) In the time domain, the response to a step function input disturbance is characterized by the time constant T. (b) In the frequency domain, the spectrum is characterized by the "corner frequency" ω_c. The term "corner frequency" may be expressed either as a frequency (Hz) or angular velocity (rad/sec); the context will make it clear which is intended.

6   SINUSOIDAL SIGNALS

a. A given linear process is completely characterized by its *frequency spectrum,* that is, its sinusoidal behavior at all frequencies. The behavior at a single frequency is specified by two quantities, the output–input amplitude ratio (*relative magnitude,* or simply *magnitude*) and the relative phase angle (*phase*); thus the frequency spectrum will consist of two parts, the (relative) magnitude and the phase. The frequency of the disturbing sinusoid may be given as a frequency (f in Hz) or as an angular velocity (ω in radians per second) where $\omega = 2\pi f$. Both measures of frequency will appear in later discussions since each has certain advantages when used in specific contexts.

b. The first-order process whose frequency spectrum is sketched here has a time constant of 0.5 sec. It is thus characterized in the time domain by the exponential response to a step function disturbance shown in (a).

c. The two portions of the frequency spectrum are shown in (b). It may be assumed that these two curves have been obtained by measurement or calculation. In either case, they have the properties previously described for a first-order flow process:

(1) A magnitude of unity at very low or "zero" frequency that then decreases approaching zero at very high values of ω.

(2) A phase that is zero at "zero" frequency and increases in a negative sense toward the value of $-\pi/2$ or $-90°$ at very high frequency.

d. The term *zero frequency,* although somewhat paradoxical, is used to denote the limit as the stimulus frequency is made smaller and smaller. Obviously, if the frequency is made zero there is no longer a sinusoidal stimulus, but instead a steady-state response to a constant disturbance. Thus, in the limit, zero frequency and a constant maintained disturbance are synonymous, and the spectrum should exhibit zero frequency relations that are the same as those developed for the constant steady state.

e. The process is further characterized in the frequency domain by the *corner frequency* (f_c or ω_c at which $\angle G = -45°$). Note that $\omega_c = 2\pi f_c = 1/T$. This relation holds for all first-order processes, and provides a direct relation between the behavior in the time and frequency domains.

f. Subsequent discussion will show that the time domain and the frequency domain provide two quite different, but entirely equivalent, viewpoints in the study of linear system behavior.

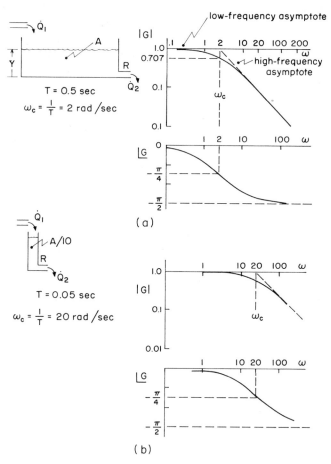

Fig. 6-7. Frequency spectra for two first-order processes drawn in logarithmic coordinates. The spectrum for a process with the time constant of 0.5 sec as in (a) is identical in form with that for a time constant of 0.05 sec as drawn in (b). The only difference in the two spectra is in their location along the ω-axis.

6 SINUSOIDAL SIGNALS

a.   The frequency spectrum shown in the previous figure was drawn with linear scales for all axes. It has been redrawn here with logarithmic coordinates; the two horizontal scales are logarithmic, as is the vertical scale for $|G|$. A linear scale has been retained for $\angle G$. Although the magnitude and phase components may appear to be quite similar to those drawn with linear coordinates, there are some important differences.

b.   Inasmuch as the ω scale is logarithmic, the value $\omega = 0$ does not appear; it is at an infinite distance off to the left. The logarithmic scale for ω permits one to select that range of values of ω of interest for a specific system, and in effect magnify it so that this range occupies the center of the plot. A similar comment applies to the scale for $|G|$.

c.   At low values of ω, the magnitude curve, $|G|$, approaches the value 1.0 asymptotically. At large values of ω, $|G|$ also approaches an asymptote, a line with the slope of -1. In this respect the logarithmic plot appears to differ markedly from the one with linear coordinates. The difference is a matter of presentation only, as examination of the two spectra will show.

d.   One distinct advantage of the logarithmic plot is that $|G|$ is quite well described by its low- and high-frequency asymptotes, which are straight lines that intersect at ω_c. If equal scales are used for the horizontal and vertical axes, then the high-frequency asymptote has a slope of -1. For many purposes the asymptotes provide a satisfactory approximation to the actual spectrum, and it is only in the vicinity of ω_c that there may be an appreciable error.

e.   The two spectra shown in this figure have identical shapes; the only distinction between them lies in their location along the ω axis. It may be shown that the spectrum of a first-order process has an invariant shape when plotted in logarithmic coordinates, and it is completely defined by its corner frequency ω_c and its zero-frequency magnitude. These are termed Bode plots after their originator (Bode, 1945).

f.   In the several spectra shown thus far, the magnitude decreases as ω increases. For this reason, the magnitude portion of the spectrum is sometimes termed the *attenuation curve,* because the magnitude is everywhere less than its value at zero frequency. However, in some systems (see Fig. 6-13) the magnitude actually increases above its zero frequency value for some intermediate range of ω, and the system may be said to exhibit some resonance phenomena. Even when this is the case, the magnitude curve is still sometimes loosely termed the attenuation curve.

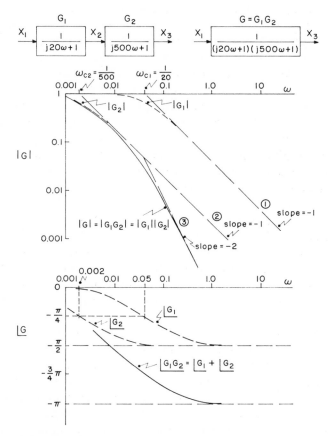

Fig. 6-8. Frequency spectrum of a second-order system formed from the two first-order spectra of the individual components. The magnitude spectrum is obtained by multiplying the magnitude components of the individual processes. This becomes addition when logarithmic coordinates are used, as in this figure.

6   SINUSOIDAL SIGNALS

a.   Part of the utility of the frequency spectrum viewpoint can be seen by considering the two compartments, the block diagram and transfer functions of which are shown here. Assume the first process in the sequence to be disturbed sinusoidally so that its output is sinusoidal, and it becomes the input to the second process. At $\omega = 0.1$ rad/sec, $|G_1|$ is 0.445 and $|G_2|$ is 0.02 so that $|G| = 0.0089$. In a somewhat similar manner the phase angle for the entire system at any angular velocity is the sum of the phase angles contributed by the two component processes. Thus, each process makes its own contribution to the spectrum for the system, although these individual contributions will change with ω. At very low angular velocities, only G_2 makes an appreciable contribution, and the process represented by G_1, having the smaller time constant, has little effect upon the spectrum.

b.   With the use of logarithmic coordinates, the calculation of the overall spectrum for a sequence of noninteracting processes is greatly expedited. If the individual processes have magnitude curves represented by the asymptotes marked ① and ②, then the system curve is obtained by adding ordinates, and it will have a high frequency slope of −2. If one uses only the asymptotes as approximations to the actual curves, then the asymptotes for $|G|$ can be drawn directly from the asymptotes of the component processes. As before, the asymptotes depart from the actual curve in a significant manner only in the vicinity of the corner angular velocities.

c.   The complete phase curve is constructed from the component curves by adding the contributions from each, point by point. Although it is possible to construct asymptotes that approximate the phase curves, these do not prove to be as good an approximation to the actual curve as is the case for the magnitude curves. Note that the total phase approaches the value of −180° at high frequency, which is −90° times the number of first-order processes, as noted earlier.

d.   The procedure for constructing the spectrum for a chain of processes as given in this figure can be extended to any number of compartments, providing that the processes are noninteracting so that the transfer functions can be multiplied. In the case of interacting processes, it is not possible to develop a transfer function for each component, independent of the other components, and although the concept of the frequency spectrum is still valid, it would have to be computed by means of a more extended procedure.

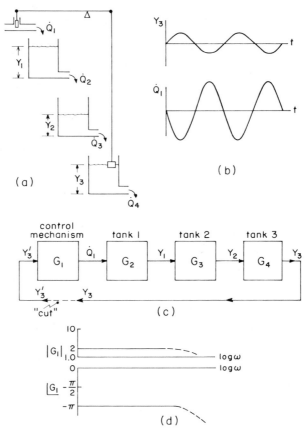

Fig. 6-9. Forced sinusoidal study of a third-order feedback system. (a) The physical system. (b) The control mechanism (float, linkage, and valve) reverses the sign of a signal impressed at Y_3 so that $\dot{Q}_1$ exhibits a phase lag of 180°. The relative magnitude of these two variables will depend upon the linkage and valve characteristics. (c) Block diagram of the system with a "cut" made at Y_3 for test purposes. (d) Frequency spectrum of the (ideal) control mechanism having no inertia.

a. The use of test functions that are sinusoids, and the representation of a system by means of its frequency spectrum, have important applications in the study of feedback systems. Although this conceptual tool is applicable only to linear systems, for which a sinusoidal input produces a sinusoidal output, it becomes possible to visualize certain relations that otherwise are quite obscure.

b. The comparison between open- and closed-loop dynamics in the last chapter should cause one to expect some corresponding changes in the respective frequency spectra. This is indeed the case, but for the purposes of this section the discussion is restricted to open-loop spectra. Thus the feedback loop will be assumed to be "cut" in the manner shown in Fig. 4-6 and again in (c) of this figure.

c. The third-order feedback system is represented by the block diagram in which the variables Y_1 and Y_2 appear. The choice of these specific variables is quite arbitrary, and instead of these quantities one could use the flow rates $\dot{Q}_2$ and $\dot{Q}_3$.

d. Frequency spectra for the tanks have already been described, but not that for the control mechanism, which is assumed to have no dynamics as discussed earlier. However, it may have a steady-state gain of other than unity, depending upon the position of the fulcrum. For this figure the gain is taken as two. The phase shift is $-\pi$ at all frequencies; this is a statement of the fact that $\dot{Q}_1$ changes in the opposite sense to that of Y_3.

e. Although the spectrum for the control mechanism alone (d) shows a constant magnitude and phase for all frequencies, it must be recognized that if the frequency were high enough, the assumption of negligible mass in the control linkage would no longer be valid. G_1 would then change, as suggested by the dotted portion of its spectrum, but presumably this would occur at high values of ω and is of no significance.

f. The "open-loop" frequency spectrum could be obtained by impressing a sinusoidal signal at Y_3' and observing the response at Y_3, or it could be constructed by combining the spectra of the individual components. The control mechanism itself has the effect of increasing the magnitude at all frequencies by a factor of two, and of adding an additional phase of $-180°$.

g. It is usually possible to open (cut) the loop in a technological system as suggested in (c) and still retain the normal characteristics of all components. However, this is much more difficult to do in a physiological system; further discussion of this problem appears in Chapter 8.

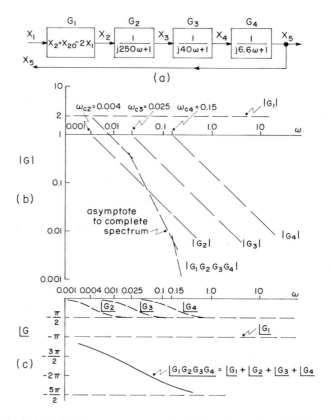

Fig. 6-10. Open-loop frequency spectrum of a third-order system including the control mechanism G_1, which has a magnitude at all frequencies of two, and a constant phase angle of $-180°$. The magnitude portion of the spectrum is represented by its asymptotes only, which serve as adequate approximations to the true curve except possibly in the region of the corner frequencies.

a. The open-loop spectrum for the entire system of Fig. 6-9 is constructed by adding the contributions from the four components in the manner previously described. This is done in this figure, where the amplitude asymptotes and phase curves for each component are shown individually and together. The ease with which the spectra of individual components may be combined is a prime reason for the widespread use of frequency spectra.

b. Although the individual tanks have gains of unity, the control mechanism may have a gain that is appreciably greater. It is here shown as two, and this then becomes the (open-) loop gain. Note that the successive tanks greatly attenuate the signal transmission so that at, say, $\omega = 0.1$ rad/sec, the transmission around the loop has fallen to nearly 0.01. This means that a sine wave of any amplitude used as a disturbing signal at 0.1 rad/sec is attenuated by a factor of almost 100 after transmission through the four components.

c. At very low values of ω the phase angle is $-\pi$, contributed by the control mechanism. This increases to a value of $-5\pi/2$ at high values of ω.

d. These curves have been drawn for corner frequencies (angular velocities) of 0.004, 0.025, and 0.15 rad/sec. This corresponds to $T = 250$, 40, and 6.6 sec. If, however, the time constants had been 25, 4, and 0.66 sec, the entire spectrum would be moved to the right by a factor of 10. There would be no changes in the *shape* of the spectrum, only in its position along the ω-axis. Both the attenuation and the phase portions move the same amount.

e. Many aspects of system behavior depend only upon the shape of the spectrum. The previous section has shown that if all time constants are changed by the same *factor* (thus preserving the time constant ratios) the shape of the spectrum is unchanged, and only its location along the ω-axis is affected. Furthermore, a change in loop gain will raise (or lower) the magnitude portion of the spectrum without changing its shape, and will have no effect on the phase.

f. It is also possible to develop certain relations between the slope of the magnitude curve and the phase. Specifically, if the slope becomes constant (n) over a sufficiently broad range of ω, then the phase approaches the constant value of $-n\pi/2$. This may be seen in the figure, where at high frequencies the slope becomes -3, and the phase of the three compartments $-3\pi/2$. The control mechanism is in a sense an exception to this rule.

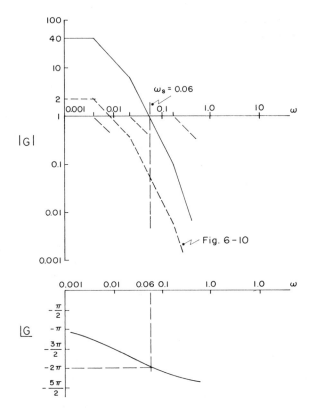

Fig. 6-11. Open-loop frequency spectrum of the third-order system of Fig. 6-10, showing the effect of an increase in loop gain by a factor of 20. With the resulting loop gain of 40, the phase angle for this system is -2π when $|G| = 1$. Note that the phase angles are not affected by this change in gain.

6   SINUSOIDAL SIGNALS

a. In technological regulators the loop gain is a readily adjustable parameter, but this is probably not true of a homeostat. However, because the loop gain has an important bearing on the performance of a feedback system, it must be examined in greater detail.

b. The frequency spectrum varies in a very simple way as the loop gain is changed. Loop gain, or loop transmission at "zero" frequency, is the intercept of the low frequency asymptote with the vertical axis. Although $\omega = 0$ does not appear on the logarithmic scale, it is assumed that the magnitude of the transmission does not change if the input were to be selected at lower frequencies than those shown. In Figs. 6-7 and 6-8 the low frequency transmission approaches unity, and the loop gain is assumed to be the same. In Fig. 6-10 the loop gain is two.

c. In this figure, the spectrum is drawn with a loop gain of 40, obtained by changing the fulcrum position. This spectrum is for the same system as that in Fig. 6-10; the time constants and the phase angles are identical.

d. Examination of these two spectra will show that the only effect of a change in gain is to raise or lower the magnitude curve by a corresponding amount. Thus, comparing the two spectra in this figure, the increase in gain by a factor of 20 has raised the entire magnitude curve by log 20. The phase angle spectrum remains unchanged.

e. The increased gain shown in this figure was selected with a view to the discussion of stability in the next chapter. Note that the gain was chosen so that the transmission drops to unity when the phase angle is -2π. This happens to occur at approximately 0.06 rad/sec. Such a point, defined by a transmission of unity and a phase of -2π, is termed a *critical point* and its significance is discussed later.

f. A recapitulation of some key ideas emphasizes that frequency spectra depict the results of a specific series of "what if" experiments. What if the feedback system were cut (as in Fig. 6-9) and then stimulated sinusoidally over a range of frequencies? The consequence would be a sinusoidal response from the other end of the cut at the same angular velocity but with different amplitude and phase. The frequency spectrum depicts the results of such tests performed at all frequencies.

g. The spectra discussed in this and the previous sections are termed *transmission spectra* because they describe the transmission properties of a physical system for sinusoidal signals of all frequencies.

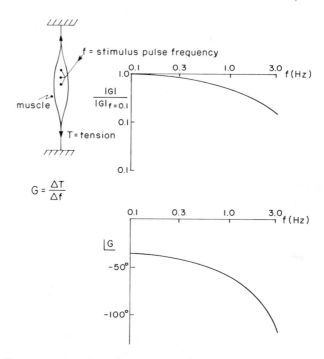

Fig. 6-12. Frequency spectrum of a nerve-muscle preparation (triceps surae in the cat) obtained experimentally under isometric conditions (Partridge, 1965). The neural pulse frequency in the motor nerve was modulated sinusoidally at the stimulus frequency f. The magnitude curve has been normalized to its value at 0.1 Hz. Note that the frequency scale is in hertz rather than radians per second.

a. The frequency spectrum has been measured by modulating the stimulating pulse frequency in the motor nerve. Two frequencies must be distinguished: (1) that of the neural pulse train, the value of which varied between 5 and 30 pps, and (2) the modulation (stimulus) frequency, which ranged from 0.1 to 3.0 Hz. Sinusoidal modulation of the pulse train caused the neural pulse grouping, or density, to take on a sinusoidal distribution.

b. The transfer function is given by the ratio

$$G = \frac{\text{change in isometric tension}}{\text{change in neural pulse frequency}} \frac{\text{(gm)}}{\text{(pulses/sec}^2)}.$$

The spectrum is plotted from the magnitude of this ratio, and the phase shift as measured from the neural signal to muscle tension. Inasmuch as G has the dimensions indicated, it is convenient to normalize the magnitude, setting $|G| = 1$ at 0.1 Hz. Presumably $|G|$ remains at unity for all lower frequencies.

c. Justification for describing the muscle by means of a frequency spectrum rests on the demonstration that a sinusoidal response is obtained for a sinusoidal stimulus, and that over some range of stimulus conditions linear behavior is observed. The experiments seem to warrant this linear model, at least as a good approximation.

d. Among the factors causing departure from linearity is the fact that the input signal (a sinusoidally modulated pulse train) cannot be a precise representation of a sinusoid. It becomes a better representation as the average neural frequency is increased, and the percentage modulation is decreased (Partridge, 1966; Borsellino *et al.,* 1961).

e. Although the magnitude curve would seem to suggest a single-compartment process, the spectrum should be available for about three decades in order to safely reach such a conclusion.

f. The low frequency phase of approximately $-35°$ might be attributed to transport lag, a process not shown in previous figures, but discussed more fully in Chapter 7. The fact that the phase at high frequency exceeds $-90°$ may be attributed to some combination of transport lag and additional compartments, but a broader spectrum is required to distinguish between these two.

g. The frequency spectrum might be used to distinguish between the fast and slow muscles. The spectrum in this figure is a composite one for all the muscles in triceps surae. Tests on gastrocnemius and soleus muscles individually seem to reveal significant differences in their spectra (Partridge, 1965).

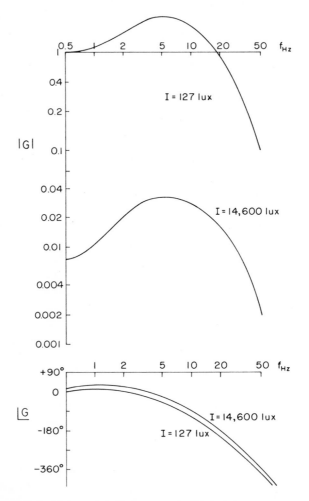

Fig. 6-13. Frequency spectra of the ERG potential measured on the eye of a wolf spider. From R. DeVoe. *Gen. Physiol.* **50**, 1967, 2008, by permission.

6 SINUSOIDAL SIGNALS

a. Spectra provide a means for describing and studying the dynamic behavior of a variety of receptor and effector organs as well as more extensive physiological systems. This figure shows the experimental results obtained from a photoreceptor and the consequences of adaptation.

b. The ERG (electroretinogram) is the electric potential measured between an electrode placed on the cornea and an indifferent electrode, in response to sinusoidal modulation of light intensity at a given wavelength. The experiments show that the ERG potentials are sinusoids over the frequencies studied, and that over a range of stimulus amplitudes, these potentials are linearly related to the stimulus intensity. The magnitude portion of the spectrum is given as relative sensitivity, obtained from the stimulus amplitude required to produce a measured response of 10 μV at all frequencies.

c. The figure shows spectra for two different average illumination levels, the eye having adapted to each level before experiments were made. The sensitivity at the higher level is substantially reduced, and there is a slight, but reasonably constant, difference in phase for the two levels. The shape of the spectrum is apparently independent of the average illumination level, and the higher level results only in a gain change.

d. This spectrum exhibits a peak in the magnitude at about 5 Hz, a feature that has not appeared in any of the previously described spectra. It is of interest to note that a similar peak has been described in the spectra of other eyes. Although the physical processes giving rise to this peak are not known, further discussion of this resonance phenomena appears later.

e. The existence of a peak does not imply a specific physical process, but it does rule out a large number of processes that are known to be incapable of producing such a response. In particular, a series of compartments (without feedback) such as described in earlier chapters is incapable of producing a spectrum with a peak in the magnitude.

f. For incremental stimuli, it has been shown that the receptor may be represented by a linear model (DeVoe, 1963). This conclusion is reached by finding that (1) sinusoidal stimuli evoke sinusoidal responses, (2) superposition holds, (3) response to an incremental step stimulus can be predicted from the measured frequency spectrum, and (4) a linear relation holds between flash intensity and response amplitude.

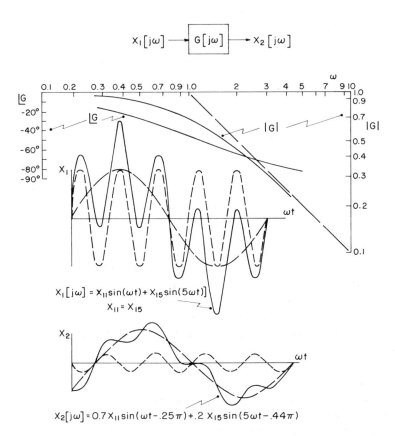

Fig. 6-14. Transmission of a periodic signal through a first-order process. The two sinusoidal components of the input signal $X_1[j\omega]$ are each modified in magnitude and phase by the process and then combined to form the output signal $X_2[j\omega]$.

6   SINUSOIDAL SIGNALS

    a. Almost any physical process will alter the shape of a periodic signal transmitted through it, and this change is clearly illustrated by considering sinusoidal signals and combinations of sinusoids. The process shown here is completely defined by its frequency spectrum, shown at the top of the figure, for which the corner frequency is 1 rad/sec.

    b. The periodic signal $X_1[j\omega]$ consists of a fundamental of angular velocity 1 rad/sec, and a fifth harmonic of angular velocity 5 rad/sec; the two components are of equal amplitude. The output signal $X_2[j\omega]$ is found by treating the two components individually. The fundamental is attenuated by the factor 0.7 and its phase is shifted by $-45°$. The fifth harmonic is attenuated by 0.2 and has a phase of 79°, or 0.44 π radian. The output is simply the sum of these two components.

    c. The above calculation makes use of the principle of superposition which is only applicable to linear processes. In effect, each sinusoidal component is treated separately, and the output signal is the sum of the components transmitted at each frequency. It is important to note that in this (linear) process, no sinusoidal components appear in $X_2[j\omega]$ that are not originally present in $X_1[j\omega]$. The converse is not true and sinusoidal components in $X_1[j\omega]$ may not appear in $X_2[j\omega]$ if the process attenuates those angular velocities to vanishingly small values.

    d. This figure shows the construction of a periodic wave as the sum of sinusoids, the angular velocities of which are integral multiples of each other. This might be termed *Fourier synthesis.* The converse, *Fourier analysis,* is the process by which a periodic signal may be resolved into a series consisting of a fundamental sinusoid having the same period as the original periodic wave, and a number of harmonic sinusoids having angular velocities that are integral multiples of the fundamental. Such an analysis can be performed for practically any periodic function; there may be a finite or an infinite number of sinusoidal components in the series. The fact that a periodic waveform can be described as the sum of a number of sinusoids points to the fundamental nature of the sine wave.

    e. Although the fundamental and fifth harmonic of X_1 in this figure are depicted as passing through zero at the same instant, this is distinctly a special case. More generally, the harmonics are found to be shifted along the time axis with respect to the fundamental, and the several amplitudes need not be equal. Thus, by summing a relatively small number of sinusoids, it is possible to obtain a periodic signal that bears little similarity to a sinusoid.

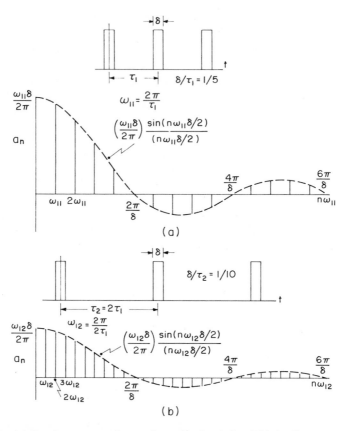

Fig. 6-15. (a) Fourier spectrum for a pulse train of period τ_1. This is a line spectrum whose components have frequencies that are multiples of ω_{11}. (b) Same as in (a), but the pulse train has a period of $2\tau_1$, and the line spectrum has components that are twice as close together.

a. The Fourier transform is a means of treating signals that are not periodic by means of sinusoids, and this concept can be developed from the Fourier series. The pulses shown in (a) are assumed to be part of an infinite pulse train of period τ_1 for which the fundamental angular velocity is ω_{11}. The Fourier series for this pulse train is represented by the spectrum shown in (a). Each vertical line represents a sinusoidal component in the series; the height of the line being proportional to the amplitude of the component. Those terms having a negative amplitude denote sinusoids with a phase shift of 180° with respect to the fundamental.

b. For this example, all integral multiples of the fundamental are present, with the exception of those having an angular velocity that is a multiple of $2\pi/\delta$. There are an infinite number of harmonics in this series, although the amplitude of the higher harmonics approaches zero.

c. In the second example (b), the period is $2\tau_1$, the fundamental is π/τ_1, and the harmonics are twice as close together as in (a). Furthermore, the amplitudes of all the terms are reduced to half that in (a).

d. If the period of the pulse train is again doubled, there will be a similar effect upon the spectrum, and this process can be continued indefinitely. In the limit the period becomes infinitely long, or what amounts to the same thing, there remains only a single pulse, and the signal is no longer periodic. For this limiting case, sinusoids of all angular velocities are present in the spectrum, although the amplitude of each component approaches zero. The spectrum is now said to be *continuous,* in contrast to the spectra shown in this figure, which are *line spectra.* In both cases there will generally be an infinite number of components. However, line spectra are distinguished by the fact that there is a finite separation between the components, whereas in a continuous spectrum all frequencies are present.

e. This figure depicts the transition from a periodic signal with its line spectrum to a nonperiodic signal with its continuous spectrum. The transformation of a nonperiodic time signal into its spectrum is termed the direct *Fourier transform,* and well-established mathematical techniques are available for doing this. The spectrum so obtained contains both amplitude and phase information for all frequencies, and it becomes possible to represent an arbitrary time signal by combining continuous sinusoids according to the "rules" contained in the spectrum.

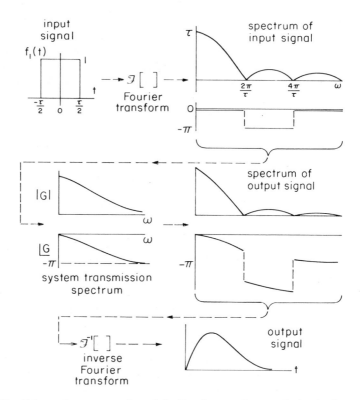

Fig. 6-16. Schematic representation of the Fourier transform analysis of a linear system. The input signal $f_i(t)$ is transformed into its continuous spectrum (magnitude and phase), which is then used as the input to the physical system whose transmission spectrum $G[j\omega]$ is known. The spectrum of the output signal is obtained by modifying each sinusoidal component in the input signal by the transmission characteristics of the system at that frequency. The output signal is found by summing these individual sinusoidal components, a process known as the inverse Fourier transform.

6   SINUSOIDAL SIGNALS

a. The previous section indicated that a nonperiodic signal could be represented by an infinite number of sinusoids making up what is termed its spectrum. The operation of calculating the spectrum from the time signal is termed the *direct Fourier transform,* and the converse, deriving the time signal from a spectrum, is termed the *inverse Fourier transform.* This figure outlines the steps by which the transforms are used to calculate system behavior for a pulse input signal. The pulse is chosen for convenience; other input signals would be treated in a similar manner (Jennison, 1961).

b. The Fourier transform yields both the magnitude and phase portions of a spectrum having the following significance. If the sinusoids appearing in the spectrum were added, each with its appropriate amplitude and phase, the result would be the original time signal. Note that the sinusoids appearing in the spectrum are continuous signals, existing for all time, but their sum is the original pulse extending over only a finite time period, and zero at all other times.

c. Having transformed the temporal input signal into a spectrum of sinusoids, it is possible to calculate how each component is transmitted by the system under study. That is, the system will transmit and modify each of the component sinusoids in the manner given by the transmission spectrum of the system. The spectrum of the output signal is thus the input spectrum modified by the transmission spectrum.

d. The final step is the inverse Fourier transform in which the output signal spectrum is transformed to give the output signal. The Fourier transforms make use of the principle of superposition, and are therefore only applicable to linear systems. Furthermore, all these spectra are continuous and contain sinusoids of all frequencies (angular velocities). The direct and inverse Fourier transforms therefore cannot be carried out with ordinary arithmetic, but require the integral calculus in which form the transforms are normally expressed.

e. The above discussion leads to the concept that a time signal and its spectrum form a pair (the *Fourier transform pair*), each of which describes a given signal in an equivalent but quite unlike manner. This duality permits one to study system behavior in either the time or the sinusoidal domain as the need arises, and leads to many useful qualitative concepts even though precise calculations are not undertaken. It further emphasizes the wide significance of sinusoidal signals in system analysis.

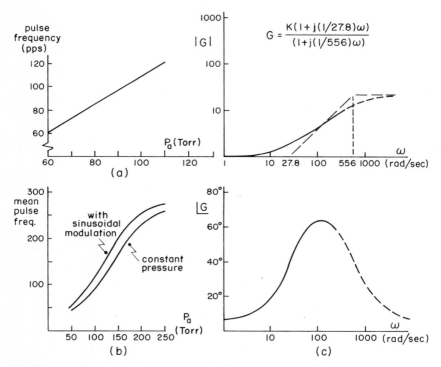

Fig. 6-17. Properties of the baroreceptor. (a) Steady-state characteristics of a fiber bundle from the rabbit baroreceptor (Bronk and Stella, 1932). (b) Steady-state characteristic for fiber bundles in the dog, with constant pressure, and with small modulation of the pressure at 2 Hz. From J. W. Spickler and P. Kezdi, *Amer. J. Physiol.* **212,** 1967, 472, by permission. (c) Bode plots of the frequency spectrum for dog (Spickler and Kezdi, 1967). Experimental points were obtained only for those portions of the spectrum shown in solid lines; the dotted portions are extrapolations calculated from the hypothetical transfer function shown in (c).

a. The baro- or pressoreceptor acts as a transducer to convert arterial pressure into appropriate neural signals. Presumably the walls of the aorta are stretched by increases in blood pressure and the resulting deformation of the nerve endings serves as an adequate stimulus for the generation of nerve impulses (a).

b. The output signal is considered to be the total impulse traffic in the hundred or so nerve fibers that originate in each receptor unit. Both recruitment of individual fibers and the increase in pulse frequency in single fibers contribute to the total biological signal.

c. The output signal is a measure not only of the mean arterial pressure, but also reflects the variation in pressure throughout the pulse wave.

d. In somewhat similar studies in the dog, the mean pulse frequency is also found to be a linear function of the pressure (b) over a range of pressures from about 100 to 200 Torr.

e. This receptor is responsive not only to the mean pressure, but also to the rate of change of pressure; that is, there is a derivative component in the output signal. This is shown in (b), where the response to a constant pressure is contrasted with that obtained when a sinusoidal modulation at 2 Hz was superimposed. If the system were linear, the addition of the sinusoidal stimulus should not affect the average impulse frequency, but the fact that it did is evidence for a nonlinearity. One might infer that the impulse rate was increased more when dP_a/dt was positive than it was decreased when the rate was negative. In any case, this nonlinearity serves to limit the usefulness of the frequency spectrum in (c)

f. In plotting the spectra, only the fundamental component in the response waveform was used; this then becomes the *describing function*.

g. The experimental points supplied only the low-frequency portions of the curves (shown solid). Physical reasoning shows that $|G|$ could not increase indefinitely, but might approach a constant value at high frequencies. The shape of the phase curve as obtained experimentally suggests the high-frequency behavior of the receptor. The transfer function shown must be considered conjectural.

h. The experimental points leading to the curves in (c) are rather widely distributed, both with regard to each animal, and also between successive animals. Averaging and normalizing were used to obtain this one curve so that it does not represent any one animal.

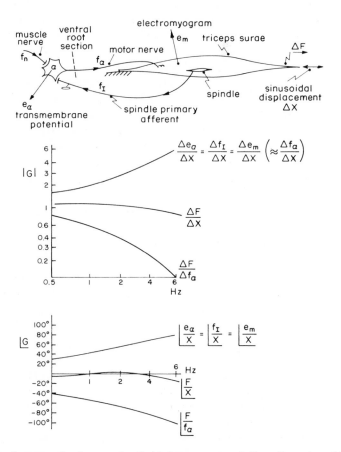

Fig. 6-18. Spectrum for the stretch reflex (triceps surae, cat). From Poppele and Terzuolo, *Science*, **159**, 1968, p. 744; copyright 1968 by the American Association for the Advancement of Science. Muscle was stretched sinusoidally by the amount ΔX. f_I is the neural signal in the spindle primary afferents. f_α is the signal in the motor nerve from the α motor neurons. The ratio $\Delta f_I / \Delta X$ is found to be almost the same, regardless of whether or not the ventral roots were cut, indicating that muscle contraction had little effect on the signal f_1.

6   SINUSOIDAL SIGNALS

    a. The stretch reflex arc (shown previously in Fig. 4–19) may be divided into three components: (1) the motor neuron pool in the spinal cord with efferent signals designated f_α, (2) the muscle innervated by the α fibers, and (3) the spindles whose primary afferent fibers synapse on the α motor neurons. As in the previous example, the neural signals f_I and f_α must be considered as the total impulse traffic in the nerve.

    b. This reflex has been tested by imposing sinusoidal changes in length, designated as ΔX. The sinusoidal response has been measured at a number of points in the reflex arc, and for as wide a range of frequencies as the experimental techniques would permit. The innervation of the γ fibers (not shown) has been assumed to remain constant.

    c. The ratio $\Delta e_\alpha/\Delta X$ designates the relative amplitude of these two sinusoids, and additional tests showed that this ratio is also very nearly equal to the ratios $\Delta f_1/\Delta X$ and $\Delta e_m/\Delta X$. The equality of these three ratios at all frequencies indicates that essentially no dynamics are introduced by the demodulation and subsequent modulation of pulse signals in the motor neuron pool.

    d. While the spectrum for $\Delta e_\alpha/\Delta X$ shows an increasing amplitude and leading phase angle with increasing frequency, that for $\Delta F/\Delta f_\alpha$ shows decreasing amplitude and lagging phase. These two spectra when combined yield the overall spectrum of $\Delta F/\Delta X$ which appears to have a remarkably constant amplitude for all frequencies shown, and practically no phase shift.

    e. On the basis of this experimental evidence, one would conclude that the spindle serves to compensate for the dynamics of the muscle. This comes about because the spindle responds to not only the stretch but also the rate of change of stretch, in a somewhat similar manner to the baroreceptor in Section 6.17.

    f. The limited range of frequencies available from this experiment makes it next to impossible to fully characterize the separate components. From other experiments with the muscle spindle, one would conclude that the phase angle probably returns to $0°$ at higher frequencies very much in the manner described in Section 6.17. While the shape of the spectrum at high frequencies may affect other properties of the reflex, the conclusions drawn above remain valid as regards compensation for lagging phase.

g. In the same vein, the phase behavior of the muscle at high frequencies cannot be estimated from this spectrum, and so the order of the system is still in doubt. Furthermore, the neural paths introduce some transportation lag (discussed subsequently) which cannot be evaluated from the data given here.

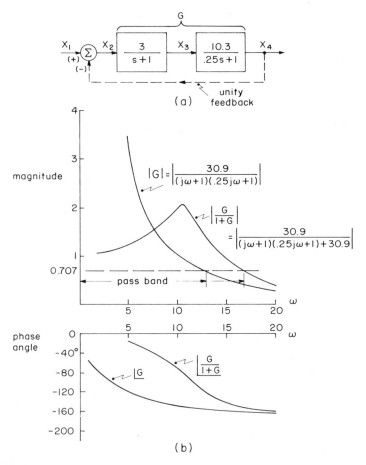

Fig. 6-19. (a) The block diagram of a second-order system. (b) The open-loop spectrum G is contrasted with the closed-loop spectrum $G/(1 + G)$. Note the transition from the Laplace variable s to the sinusoidal operator $j\omega$ when describing the sinusoidal response. In this figure linear scales have been used for the spectra.

6 SINUSOIDAL SIGNALS

a. The spectrum is also useful in depicting the effects of feedback. This may be done by comparing the spectrum of a physical system, when measured without feedback, with that obtained when the feedback is effective. The second-order system shown in (a) is to be considered as a specific example.

b. The magnitude portion of the open-loop spectrum $|G[j\omega]|$ falls monotonically toward zero as the frequency increases. From the transfer function it is seen that at $\omega = 0$ the magnitude is 30.9; this is the loop gain. The phase portion of the spectrum ranges from 0° toward 180° at large values of ω. Note that $\angle(G[j\omega])$ only approaches 180° at high frequencies but never reaches that value.

c. The closed-loop spectrum may be calculated from the closed-loop transfer function, using complex algebra, or may be found by graphical methods (Milsum, 1966). It is obviously quite different from the open-loop spectrum. At $\omega = 0$, the magnitude (gain) is 30.9/(1 + 30.9), and at approximately $\omega = 11$ rad/sec there is a resonant peak with the magnitude reaching a value of approximately two. The phase exhibits the same range of values as in the open loop (0 to 180°), but the curve has a somewhat different shape, and the change in angle occurs at higher values of ω.

d. It is convenient to define a *pass band* as that range of angular velocities over which the magnitude of the transmission exceeds 0.707. The number 0.707 is rather arbitrary, but related to the energy transmitted by a sinusoidal signal. The pass band for the closed-loop spectrum is obviously greater than that of the open loop, having increased from 13 to about 17 rad/sec, and over a large portion of this range the transmission of the closed loop appreciably exceeds that of the open loop.

e. The system gain has been reduced from 30.9 to 0.97 with the closed loop so that X_4 more closely follows X_1 at low angular velocities, should that be desired. At the same time, the sensitivity of the system to high-frequency disturbances will be greatly reduced because of the broader pass band.

f. The existence of a peak in the spectrum correlates quite well with the appearance of complex roots in the s-plane and damped sinusoidal transients following a disturbance. Furthermore, the height of the peak is inversely proportional to the damping per cycle, and the angular velocity at which this peak occurs is approximately the same as that in the transient response. These relationships between the spectrum and the transient response are quite precise for a second-order system, and although they become somewhat diffuse in systems of higher order are still helpful guides.

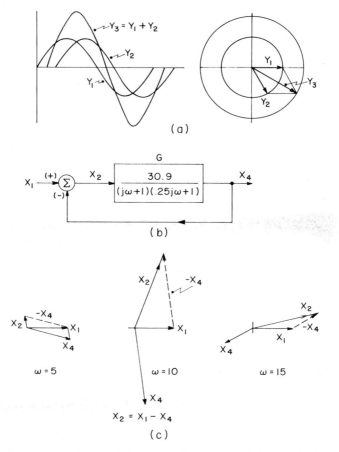

(a)

(b)

(c)

Fig. 6-20. Phase and amplitude relations in the forced sinusoidal study of a feedback system. (a) Addition of two sinusoids of the same frequency by vector addition of the phasors. (b) Block diagram of the feedback system in Fig. 6-19. (c) Phasor relations for $\omega = 5$, 10, 15 rad/sec.

a. The previous section considered the case in which unity feedback, when added to a two-compartment system, produced a spectrum having an appreciable resonant peak. This figure pertains to exactly the same system, and the following discussion attempts to relate that peak to the physical aspects of the system.

b. In broad terms, the peak is related to the relative amplitude and phase of the several sinusoidal signals within the system, and to discuss this relation, it is necessary to consider the addition and subtraction of sinusoidal quantities. The addition of two sinusoids is seen in (a). On the left, this is carried out by adding the magnitudes of the two components at each instant of time; the sum being the sinusoid shown. On the right, this addition is performed by phasor addition, using the parallelogram law. The resultant phasor represents the same sinusoid as was found by point-by-point addition.

c. For the feedback system shown in (b), one can write $X_2[j\omega] = X_1[j\omega] - X_4[j\omega]$ for the summing point. This subtraction can be turned into an addition by writing $X_2 = X_1 + (-X_4)$; that is, we add the negative of X_4 as shown in (c) for three different values of ω. Note that X_1 is kept the same for all angular velocities.

d. With $\omega = 5$ rad/sec, it will be seen that the magnitude of the loop transmission $|G|$ is considerably larger than one, and the relative phase of G is a little more than $-90°$. At $\omega = 10$ rad/sec, $|G|$ has dropped to about unity, and the phase is approaching $-180°$. At the angular velocity of 15 rad/sec, the magnitude of the loop transmission has fallen below unity, and the phase of G is even closer to $-180°$. Thus, the changes in both magnitude and phase of the loop transmission are such that the closed-loop transmission (X_4/X_1) has a magnitude in the vicinity of 11 rad/sec that is larger than it is for both lower and higher values of ω.

e. These relations might also be described by stating that under sinusoidal forcing, the summing point does not perform a simple arithmetic subtraction process but rather a phasor subtraction. The relative phase and magnitude of X_1 and X_4 are therefore both significant.

f. In a regulating system, the input X_1 may well be a constant, so that the closed-loop sinusoidal transmission X_4/X_1 may be of little significance. However, with a disturbance appearing at some other point in the system, the closed-loop transmission from that point to the regulated variable would be of interest. It could be found by the same general methods as described here.

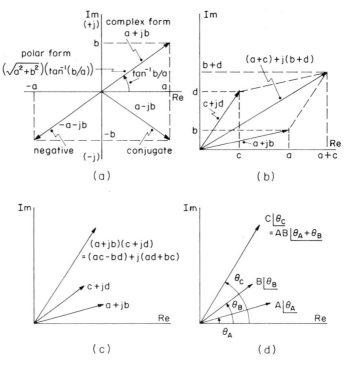

Fig. 6-21. Algebraic operations with complex numbers. (a) Phasor in its complex and polar form. (b) Addition of two complex numbers and the parallelogram law for addition. (c) Multiplication of two complex numbers. (d) Multiplication in polar form. Division is the inverse of polar multiplication, or can be carried out by rationalizing the denominator, namely,

$$\frac{a+jb}{c+jd} = \frac{(a+jb)(c-jd)}{(c+jd)(c-jd)} = \frac{(ac+bd)+j(bc-ad)}{c^2+d^2}.$$

6   SINUSOIDAL SIGNALS

a. It is sometimes necessary to calculate the spectrum of a given system directly from the differential equation or the transfer function. This would be the case, for instance, if the system contained interacting components, or in calculating the closed-loop spectrum from the open-loop functions. In such circumstances, the addition of individual Bode plots is not possible, and recourse to algebraic or other graphical methods is necessary. The algebraic method of calculating spectra is treated in this section.

b. In Section 6.3 it was simply stated that the behavior under forced sinusoidal oscillation could be obtained from the transfer function by substitution of $j\omega$ for s. This rule was given without proof, but the interested reader will find the matter discussed by Milsum (1966, p. 157), and by Johnson (1967, p. 154).

c. The application of this rule to the transfer function $G = X_2/X_1 = K/(T_1s + 1)(T_2s + 1)$ results in $G[j\omega] = K/(jT_1\omega + 1)(jT_2\omega + 1)$. On the other hand, the sinusoidal solution could be obtained from a differential equation, $a_2\ d^2X/dt^2 + a_1\ dX/dt + a_0X = Y \sin \omega t,$ by writing $j\omega$ for d/dt so that the sinusoidal transfer function becomes $X[j\omega]/Y[j\omega] = 1/(a_2(j\omega)^2 + a_1(j\omega) + a_0) = 1/(a_0 - a_2(\omega)^2 + ja_1\omega)$. This complex quantity is then calculated for each value of ω to give the corresponding magnitude and phase for the spectrum.

d. The calculation of the spectrum for a physical system requires algebraic operations on sinusoids (or their phasors) all of which have the same angular velocity. Thus, all the phasors in any one diagram rotate at the same velocity so that their relative positions remain unchanged with time. A given calculation will then be repeated for as many different values of ω as is desired.

e. A phasor may be written in either complex or polar form, as in (a); the spectrum is obtained by plotting the polar form, which gives the magnitude and phase directly. In general, the algebra of complex numbers follows the same operational laws as those used with real numbers, with the addition of the relation $\sqrt{-1} = j$, or $j^2 = -1$. The algebraic operations may be carried out in either complex or polar form, as is suggested in the figure.

f. The sum or difference of two sinusoids of the same angular velocity always yields another sinusoid of the same velocity. The algebraic operations shown in this figure are such as to give the output sinusoid when the transfer function is specified and the angular velocity given.

The analysis of system behavior is usually equated to the calculation of the time response for a given input or disturbance. The discussion in this and the previous chapters will have shown that the analysis can be carried out in two quite different manners. In Chapter 3 the differential equation, or the transfer function, was the computational tool, using either a classical analysis of the differential equation, or the Laplace transform method, suggested but not described in this volume. Both classical and Laplace methods lean heavily upon the poles and zeros of the transfer function, the locations of which serve to define the general form of the solution. Another type of analysis, that employing the frequency spectrum and the Fourier transform, has been described very briefly in this chapter. As might be expected, there is a close, well-defined relation between these two analytical techniques, and one can pass from the time domain to the frequency domain with suitable mathematical methods. More explicitly, having given the frequency spectrum of both the system and the disturbance, it is possible to calculate the time response. Furthermore, with the disturbance and time response given, one can calculate the frequency spectrum of the system.

These two types of analysis (time domain and frequency domain) serve to contrast two types of problem encountered in the engineering literature. On the one hand are those problems concerned with the dynamic response to arbitrary disturbances; these are said to be the *free* (or *natural*) *vibrations* of the system. As used here, the term vibration includes the exponential transient as a limiting case. The natural vibrations or *modes* are those responses evoked whenever the system is disturbed. In contrast, if the system is forced to oscillate sinusoidally the response is termed a *forced vibration*. Since both the free and the forced vibration represent the behavior of the same system, albeit under quite different stimulus conditions, it is to be expected that there should be a close connection between the free and forced oscillatory behavior. This is provided explicitly by the Fourier transform.

For a single compartment, the magnitude and phase components of the spectrum have invariant shapes, even though their horizontal positions vary with the corner frequency, and the vertical position of the magnitude curve varies with the system gain. From this invariance of shape, it follows that if one of these components is given alone it is possible to construct the other curve, and this can be done for a system containing any number of components. However, this intimate relation between the magnitude and phase portions of the spectrum holds for systems of any complexity, providing only that the components are processes similar to the compartments used in the previous examples. These are termed *minimum phase systems,* which means for a magnitude curve of given shape, there is always associated with it a minimum phase having the characteristics previously described. However, there are processes for which this magnitude-phase relation does not apply, and which have a greater phase than shown in the previous figures. These are termed *nonminimum phase systems,* for which the phase cannot be calculated from the magnitude curve alone. An example is the transportation lag described in the next chapter.

As a consequence of the above relations there follows some important conclusions for the minimum phase system. If it can be established that the system is indeed of the minimum phase type, then it is not necessary to measure both the magnitude and phase portions of the spectrum in an experimental study. It is fairly common practice to measure only the magnitude of the sinusoidal response, and to infer the phase curve from it. It should be noted, however, that this can be highly misleading, and is valid only if the minimum phase character is established by some other means. It is quite probable that a large proportion of physiological systems are *not* minimum phase, and the inference of phase characteristics from the magnitude curve is therefore invalid. Furthermore, the neglect of small phase angles can lead to erroneous conclusions in the evaluation of stability, as discussed more fully in the following chapter. There is an additional reason for making the simultaneous recording of phase and magnitude highly desirable. Sinusoidal studies are usually made with small signals to assure linear response and to avoid nonlinearities due to saturation effects. Small signals are invariably contaminated with noise, and it becomes very helpful to have both magnitude and phase measurements to assist in the interpretation of the results. In addition, having both portions of the spectrum available, one has clear evidence for the existance of phase angles larger than the minimum, if such exist. In some psychophysical experiments on vision, the observed response is subjective and only the magnitude portion of the spectrum can be measured. Deduction of the phase curve from such measurements could be highly misleading.

Several practical aspects of the frequency spectrum are worth noting. The pass band, those frequencies transmitted with magnitudes greater than 0.707, is the principal factor in determining the speed of response, and the sinusoidal components of the spectrum lying well above the pass band will have negligible effect. The sequence in which the corner frequencies appear along the ω-axis is only a matter of relative size, and bears no relation to the location of that component within the system. The spectrum thus provides a means of identifying those components that are most significant insofar as system behavior is concerned, and by the same token points to other components that can possibly be neglected in making an approximate analysis. A regulating system is said to have a *low pass spectrum* because the transmission always extends down to and includes "zero" frequency, or a constant signal. A physical system can be designed to have *band-pass* characteristics; that is, to transmit only a band of frequencies intermediate between zero and very high values. Although a band-pass process could not appear in the principal pathway of a regulating system, it might be located in an ancillary path to provide some desired dynamic characteristics to the system. A bandpass process located in the principal path would prevent regulation in the steady state.

Most of the previous discussion has tacitly assumed that the system contained a number of noninteracting components, so that the spectrum could be constructed by adding (with logarithmic coordinates) the contributions from each. When the compartments are interacting, however, an additional step is required. If there should be n interacting compartments, the differential equation (or the transfer function) is of the nth order, and to apply the above procedure, the nth degree characteristic equation must be factored. That is, the roots must be found so that the interacting compartments may be replaced by the same number of noninteracting compartments, the overall behavior of which is identical. Alternatively, the spectrum would have to be computed directly from the transfer function by means of the algebra of complex numbers (Section 6.19).

The relative ease with which a spectrum can be constructed makes it desirable to infer as much about the temporal behavior of the system as possible directly from the spectrum. Although the Fourier transform is the mechanism by which the time and sinusoidal domains are related, it is possible to infer a number of relations without actually carrying out the integration required in using the transform. Most of these relations have been stated previously, and are collected here for convenience.

1. The *"zero" frequency transmission* is the gain of the system which establishes the steady-state, quiescent relations, and is the dominant factor in system sensitivity to disturbances.

2. The *pass band* includes those frequencies that are transmitted with little attenuation, and hence are the most significant components in the response. Inasmuch as frequencies above the pass band are largely attenuated, the speed with which the system can respond to a disturbance

is limited by the frequencies in the upper edge of this band. The broader the pass band, the more rapid will be the system response.

3. *Peaks in the magnitude curve* (resonance) are indicative of low damping, and this appears in the time domain as transient oscillations that persist for extended periods of time.

4. In a general sense, the amount of the phase at, for example, ω_1 is related to the slope of the magnitude curve at all frequencies, but depends to the greatest extent upon the slope at ω_1. The greater the (negative) slope of the magnitude curve, the greater will be the lagging phase angle.

In particular, the relations between the phase and magnitude curves in the vicinity of 360° are of special significance from the standpoint of stability as described in Chapter 7. These several relations between the phase and magnitude components are further evidence of the intimate connection between them in minimum phase systems.

The application of spectral concepts has a number of important advantages, and this is especially true in the study of biological systems. First, the spectrum represents the system behavior under steady-state sinusoidal conditions. If the spectrum is measured experimentally, the observations should be made only when all transients have disappeared, and the sinusoidal response repeats itself from cycle to cycle. Thus, all initial conditions that might ordinarily affect the behavior are no longer a factor. In contrast, with a step function input, or with any other arbitrary disturbance, the character of the response may be greatly affected by the initial conditions. Repeatability of the response, as well as its interpretation, require maintenance of the same initial conditions and a knowledge of what they are.

Adaptation in some form is usually present in physiological systems, and experimental studies must be designed either so as to remove its effects, or with full knowledge of what they are. In many instances, a forced sinusoidal study can minimize or remove the effects of adaptation, in particular when these effects depend upon the average value of the stimulus or some parameter. In general, adaptation takes place over long periods of time, compared to the system time constants. The use of step function disturbances, for instance, may result in adaptive changes that do not appear over the normal response periods. A forced sinusoidal oscillation maintains a constant average value, which may be that of the unstimulated state, or which corresponds to some constant level of adaptation. The measurement of relative magnitude and phase are made only after the system has reached a sinusoidal steady state and there is no further change in the average value. This ability to specify and change the adaptive state makes sinusoidal studies an attractive approach with physiological systems.

1. The sinusoid is the fundamental periodic function with which all other periodic functions can be synthesized and described. It is also found that the sinusoid is a powerful tool in the study of system dynamics even though sinusoidal signals are not explicitly present.

2. A rotating line segment, termed a phasor, may be used to construct a sinusoid. The maximum magnitude of the sinusoid (termed its amplitude) is equal to the length of the phasor, and its phase is the instantaneous angle between the phasor and the horizontal axis.

3. A sinusoidal disturbance to a linear physical system produces sinusoidal variations at the same frequency in all the dependent variables. However, their relative amplitudes and phases are characteristic of the gains and time constants of intervening processes.

4. The ratio of output to input amplitudes is termed the relative magnitude, or attenuation, and both it and the phase of all dependent variables are functions of the disturbing frequency. In general, the relative magnitude approaches a constant value at very low frequencies, and becomes vanishingly small at high frequencies. Correspondingly, the phase is usually zero or $-180°$ at low frequencies and approaches some multiple of $-90°$ at very high frequencies.

5. A sinusoidal disturbance, used as a conceptual and experimental tool, is normally kept at an amplitude such that all perturbations are small compared to the average values of the variables concerned. In flow processes, the amplitude must be less than the average value to preclude negative, and thus nonphysical, values for the variables.

6. The frequency spectrum depicts the magnitude and phase of one process, or an entire system, as a function of frequency. Although theoretically the spectrum contains all frequencies from zero to infinity, in practice only a range of frequencies is shown, since the behavior at both lower and higher frequencies may be inferred.

7. The frequency spectrum when plotted with logarithmic coordinates (a Bode plot) reveals that all first-order processes have spectra of the same shape. The distinction between such spectra lies in the location of the corner frequency along the frequency axis.

8. The spectra of second-order and higher-order systems may frequently be constructed from the basic first-order spectrum, providing that the processes are noninteracting. The construction of such a spectrum is aided by the use of low and high frequency asymptotes for the component spectra. The actual spectrum departs from these asymptotes only in the vicinity of the several corner frequencies.

6   SINUSOIDAL SIGNALS

9. The open-loop frequency spectrum for a feedback system is developed by artificially opening the loop at some convenient point. The control mechanisms so far described have a spectrum with constant magnitude for all frequencies, and a constant phase of $-180°$ due to the sign reversal inherent in the linkage mechanism.

10. The open-loop spectrum for a third-order system reveals a rapidly falling magnitude curve, and an ultimate phase of $-450°$ ($-5\pi/2$). A basic property of linear systems is that the *slope* of the magnitude curve is related to the amount of phase shift.

11. The gain of a system determines the vertical position of the magnitude spectrum. An increase in gain simply raises the entire curve by a fixed amount, when logarithmic coordinates are used. The loop gain does not effect the phase portion of the spectrum.

12. Experimental limitations often restrict the range of frequencies that can be observed. Measurements over a frequency range of less than about three decades (1000) make it quite difficult to use a spectrum for identification of the physical processes.

13. Spectra obtained for the ERG of the wolf spider eye exhibit a resonance peak, but provide no clue as to the physical mechanisms producing it. Adaptation has a marked effect on the magnitude curve, but produces a very small change in the phase.

14. The transmission of a periodic signal through a linear process may be found by resolving the signal into its sinusoidal components (a Fourier series) and computing the transmission for each. The output signal is the sum of these transmitted sinusoids.

15. The Fourier series of a pulse train has a line spectrum whose components become denser (closer together) as the period of the pulse train is made greater. In the limit, a single pulse in the time domain, corresponding to a train of infinite period, yields a continuous spectrum. The Fourier transform is the means by which a nonperiodic signal is converted to a continuous spectrum.

16. The Fourier transform and its inverse permit one to calculate the response of a linear system to a reasonably arbitrary disturbance. This is accomplished by transforming the calculation to the frequency domain, and treating the problem as one of transmission of an infinite number of sinusoids.

17. The baroreceptor is an approximately linear device for small signals whose spectrum exhibits a leading phase angle over angular velocities of from 10 to 100 rad/sec.

18. The spectrum of the stretch reflex reveals that the spindle provides a leading phase angle that almost exactly compensates for the lagging angle contributed by the muscle, at least over a limited frequency range. Limitations on the frequency range attainable experimentally make it difficult to infer additional properties of the reflex.

19. Negative feedback, when added to a physical system which has a monotonically decreasing magnitude spectrum, is frequently found to produce a "resonant" peak in the closed-loop spectrum. In addition, negative feedback broadens the pass band.

20. The production of a resonant peak in the spectrum is associated with the appearance of complex roots of the characteristic equation. The presence of a peak may be understood physically from a consideration of the relative amplitude and phase of the signals at the summing point, and the way in which they change with frequency.

21. Spectra can be computed by combining the Bode plots of individual components, providing they are noninteracting. With interacting components, the spectrum must be calculated directly from the transfer function, or from the differential equation, using the algebra of complex numbers.

22. Sinusoidal analysis is frequently employed in the study of physiological systems because it not only helps to identify and separate the contributions of individual components, but also because it eliminates some of the disturbing effects produced by initial conditions and adaptation.

Problems

1. Calculate and plot, in logarithmic coordinates, the frequency spectrum (magnitude and phase) of $1/(j\omega + 1)$ over about three decades on the ω-axis. Use three-cycle log-log paper for the magnitude, and semilog paper for the phase. These curves, carefully plotted, may be traced to construct the spectra in the following problems.

2. Plot the separate components and the complete spectrum for each of the following functions.

(a) $G_1 = \dfrac{10}{(0.5j\omega + 1)(20j\omega + 1)}$; (b) $G_2 = \dfrac{50}{(100j\omega + 1)(45j\omega + 1)}$.

3. Plot the frequency spectrum for each of the following functions. Use the curves obtained in Problem 1 to plot the numerator and denominator separately. Note that the numerator term contributes a magnitude that increases with ω, and a phase that is positive rather than negative.

(a) $G_3 = \dfrac{j\omega + 1}{0.1j\omega + 1}$; (b) $G_4 = \dfrac{j\omega + 1}{10j\omega + 1}$.

6 SINUSOIDAL SIGNALS

4. Consider the first-order transfer function as plotted in Problem 1. Show that the slope of the high frequency asymptote is -1, and that of the low frequency asymptote is zero.

5. A second-order transfer function, the denominator of which has complex roots, will usually show a resonant peak in its spectrum. For each of the functions given below, (1) calculate the roots of the denominator, (2) plot the roots on the s-plane, and (3) calculate the frequency spectrum and make a log-log plot.

$$\text{(a)} \quad G_1 = \frac{25}{s^2 + 8s + 25}; \qquad \text{(b)} \quad G_2 = \frac{400}{s^2 + 5s + 50}.$$

6. (a) From the spectra of G_1 and G_2 plotted in Problem 2, find the angular velocity at gain crossover, where $|G| = 1$, for each transfer function.

(b) Find the gain necessary in Problem 2a to make the pass band 2 rad/sec.

7. If $G_1[j\omega]$ in Problem 2 is the open-loop transfer function, calculate and plot the closed-loop spectrum, $G_1/(1 + G_1)$.

8. For the system shown in Fig. 6-10, assume that $X_1 = 2 \sin(0.01t) + 5 \sin(0.02t)$. Find $X_5[j\omega]$.

9. Assume that the nerve-muscle spectrum in Fig. 6-12 may be approximated by a first-order transfer function. Estimate the time constant, and sketch the theoretical spectrum for comparison with the experimental one in this figure. Point out the deficiencies in this approximation of the experimental curve.

10. Estimate the values of $|G|$ and $\angle G$ from the curves in Fig. 6-19. Using this data, calculate the values of $|G/(1 + G)|$ and $\angle[G/(1 + G)]$, and compare with the values plotted in this figure. Make this comparison for $\omega = 5$, 10, and 15 rad/sec.

11. Confirm the fact that the *shape* of the frequency spectrum plotted in logarithmic coordinates depends only upon the time constant ratios, and not upon their absolute magnitudes.

12. Insofar as possible, compare the spectra shown in Figs. 6-12 and 6-18. Contrast the experimental procedures used in each case. Should the two spectra be comparable? Are there any significant differences?

References

Bode, H. W. (1945). "Network Analysis and Feedback Amplifier Design." Van Nostrand-Reinhold, Princeton, New Jersey.

Borsellino, A., Poppele, R. E., and Terzuolo, C. A. (1961). Transfer functions of the slowly adapting stretch receptor organ of crustacea. *Cold Spring Harbor Symp. Quant. Biol.* **30,** 581–586.

Bronk, D. W., and Stella, G. (1932). Afferent impulses in the carotid sinus nerve. *J. Cell. Comp. Physiol.* **1,** 113–130.

DeVoe, R. D. (1963). Linear relations between stimulus amplitudes of retinal action potentials from the eye of the wolf spider. *J. Gen. Physiol.* **47,** 13–32.

DeVoe, R. D. (1967). A nonlinear model for transient responses from light-adapted wolf spider eyes. *J. Gen. Physiol.* **50,** 1993–2030.

Jennison, R. C. (1961). "Fourier Transforms and Convolutions for the Experimentalist." Pergamon, Oxford.

Johnson, E. F. (1967). "Automatic Process Control." McGraw-Hill, New York.

Milsum, J. H. (1966). "Biological Control Systems Analysis." McGraw-Hill, New York.

Partridge, L. D. (1965). Modifications of neural output signals by muscles: A frequency response study. *J. Appl. Physiol.* **20,** 150–156.

Partridge, L. D. (1966). A possible source of nerve signal distortion arising in pulse rate encoding of signals. *J. Theor. Biol.* **11,** 257–281.

Poppele, R. E., and Terzuolo, C. A. (1968). Myotatic reflex: Its input output relations. *Science* **159,** 743–745.

Pringle, J. W. S., and Wilson, V. J. (1952). Response of a sense organ to a harmonic stimulus. *J. Exp. Biol.* **29,** 220–234.

Spickler, J. W., and Kezdi, P. (1967). Dynamic response characteristics of carotid sinus baroreceptors. *Amer. J. Physiol.* **212,** 472–476.

Stark, L., and Sherman, P. M. (1957). A servoanalytic study of consensual pupil reflex to light. *J. Neurophysiol.* **20,** 17–26.

Stoll, P. J. (1969). Respiratory system analysis based on sinusoidal variations of CO_2 in inspired air. *J. Appl. Physiol.* **27,** 389–399.

Trimmer, J. D. (1950). "Response of Physical Systems." Wiley, New York.

. . . the concept of stability, that much overburdened
word with an unstabilized definition.

BELLMAN (1953)

Introduction   7.1

The term stability has a variety of meanings and is used in a number of different ways. In everyday usage, stability frequently connotes a certain constancy,
a lack of change or fluctuation. In a more technical sense, stability may mean the
ability to withstand disturbances without disruption or permanent loss of normal
function. Thus, the application of this concept to physical and biological systems
requires a greater precision of definition than may be obvious from the above
statements.

The question of stability, or of its converse, instability, becomes an important
matter in systems with feedback, simply because all such systems *may* become
unstable and disruptive even though they are completely stable in the absence
of feedback. The engineer recognizes this potential instability in his design
of a feedback regulator, paying particular attention to system stability, and
incorporating into the design ample provisions to assure stable operation
under all foreseeable operating conditions. The homeostatic processes under
discussion here are normally stable as is evidenced by their continued existence, having acquired stability under evolutionary pressures. However, the
fact of stability does not remove the question from our field of inquiry. The mere
existence of stability in a physiological system leads to many questions as to how
that stability is attained and maintained, once the potential for instability is recognized. Although pathology may lead to instability, our interest lies more in the
margin between normal behavior and instability, as found experimentally in the
living system, on the one hand, and as predicted by the mathematical model on
the other.

The concept of potential instability has immediate and practical consequences
for the physiological system; in some instances it is observed clinically, and in
a few examples may be induced experimentally. With a given definition of stability, it has been possible to develop criteria for the existence of stability, expressed as quantitative and mathematical relations between the system parameters. These criteria therefore refer to the stability of the mathematical model,
and this test becomes an important indicator of the model's validity.

In the following sections, we propose to examine some of the stability criteria
that are applied to technological systems and assess their value in the study of
homeostatic processes.

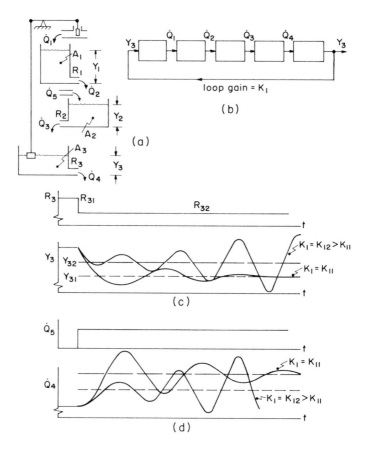

Fig. 7-1. Onset of instability in a feedback regulator as the loop gain is increased. The liquid level regulator (a) has the block diagram (b). A decrease in R_3 as in (c) results in an oscillatory response, the magnitude of which increases with time if the loop gain is sufficiently large. (d) A disturbance to $\dot{Q}_5$ produces similar results.

7 **STABILITY**

a. In the two-compartment system of Fig. 5-11, the gain is such that the response is a damped sinusoid, which appears in all the system variables. Furthermore, with an increase in the gain, as in Fig. 5-10, a large number of oscillations appear in the response.

b. With the three-compartment system shown here, additional time is required to transmit changes through the system, and this can result in a dramatic change in the behavior. The stepwise decrease in R_3 shown in (c) produces an oscillatory response which damps out if $K_1 = K_{11}$, but increases without limit if $K_1 = K_{12}$. In the latter case the system is obviously unstable.

c. With the smaller value of gain (K_{11}), the new steady state for Y_3 is Y_{31}. However, when $K_1 = K_{12}$, the indicated final value of Y_{32} is never attained simply because the oscillations continue with increasing amplitude.

d. In this unstable case, with the amplitude of oscillation becoming greater with time, there must be some point at which the amplitude ceases to increase. Either the system saturates at some point and thus limits the amplitude of oscillation, or disruption occurs and the components break down in some manner. In this liquid level regulator, the tanks would undoubtedly overflow, and thus achieve a form of saturation, but in other cases a variety of fates might await the system. The precise ultimate state of the system is not of immediate interest, but we do attach importance to the following questions. In the face of potential instability, what criteria can serve to predict it? What aspects of the system tend to promote instability? Are means available for preventing instability and promoting stable operation over all possible operating regions?

e. Part (d) of the figure shows the system response to the addition of a new input $\dot{Q}_5$. Its effect, to raise the level of Y_3, is counteracted by a reduction in $\dot{Q}_1$ through feedback. Note that an increased gain reduces the sensitivity of the system in both (c) and (d), but the figures also show the penalty that may be associated with the higher gain; namely, the potentiality of becoming unstable.

f. Instability is dependent upon the properties of the system, and is normally not a function of the specific disturbances. If the system is stable, all transients will ultimately disappear regardless of the disturbances causing them. On the other hand, any disturbance to an unstable system will initiate oscillations that increase in amplitude with time.

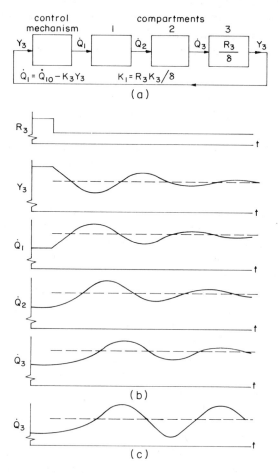

Fig. 7-2. Phase relations between the dependent variables in a feedback system. (a) Block diagram of the system in Fig. 7-1. (b) Transient behavior following a decrease in R_3, with $K_1 < K_{1c}$. (c) The case of critical gain, $K_1 = K_{1c}$.

7 STABILITY

a. This figure depicts the same system as in Fig. 7-1, but in a slightly different form, and shows relations between the transients similar to those of Section 5.11. The disturbance is again a stepwise reduction of R_3, and the ensuing behavior of the several variables has been sketched.

b. Of particular interest is the relative time or phase relations between the dependent variables, and the way in which the disturbance and the corrective signals are propagated through the system. Note that Y_3 starts to fall immediately following the change in R_3, and that $\dot{Q}_1$ also starts to change immediately, but in the opposite sense. As a matter of fact, $\dot{Q}_1$ is the mirror image of Y_3 because of the sign inversion taking place through the valve linkage. This mechanism has been assumed to introduce no dynamics into the system, in keeping with previous examples.

c. The remaining variables, $\dot{Q}_2$ and $\dot{Q}_3$, respond in similar damped oscillatory manners, but each is delayed in time by an amount proportional to the time constant of the associated compartment. These variables all have the same angular velocity, and are damped at the same rate. To make these relations somewhat clearer it has been assumed for this figure that the three time constants representing the compartments are approximately equal so that each compartment contributes the same amount to the total phase shift. This treatment of a damped oscillation as though it were a sinusoid is not quite correct, and a further refinement of this viewpoint appears necessary. With this comment in mind, it may be seen that the phase shift from $\dot{Q}_1$ to $\dot{Q}_2$ is close to 60°, and a similar relation exists for $\dot{Q}_2$ and $\dot{Q}_3$, and for $\dot{Q}_3$ and Y_3. If each of these relative phases had been 60°, the total, together with the 180° contributed by the valve mechanism would yield 360°. This is approximately the case in (c), where the $\dot{Q}_3$ reinforces Y_3, the oscillations persist, and the system is termed unstable.

d. For a given system, the loop gain will be an important factor as regards instability, and the value of K_1 on the boundary between stability and instability is termed the *critical gain,* and denoted K_{1c}. At this value of gain, the sinusoidal transmission around the loop for some angular velocity is unity, and the total phase shift is 360° or zero. Thus, a transient, once initiated, perpetuates itself, and oscillations persist.

e. A system of second order (containing only two compartments) cannot become unstable in this manner simply because the phase shift never equals 360°, although it may approach that value.

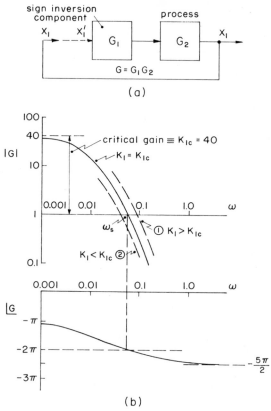

(a)

(b)

Fig. 7-3. Critical gain for a feedback system as defined by the open-loop spectrum. (a) Block diagram of the system. (b) Spectrum with $K_1 = K_{1c}$, for which $|G| = 1$ and $< G = -2\pi$ at $\omega = \omega_s$. If $K_1 > K_{1c}$, the spectrum would shift to ①, and if $K_1 < K_{1c}$, the spectrum would move to ②.

a.   The ideas introduced in the previous section can be given a more precise formulation by means of the frequency spectrum. The one in (b) is for a three-compartment system, and is the same as that shown in Fig. 6-11. This spectrum describes the open-loop behavior from 0.001 to 1.0 rad/sec, within which range $|G|$ falls from about 40 to less than 0.1, and the total phase shift $<G$ varies from $-\pi$ to $-5\pi/2$.

b.   The spectrum in solid lines has been drawn for critical gain K_{1c} so that $|G| = 1$ when $\angle G = -360°$, and this occurs at ω_s. The spectrum thus gives both a physical picture of the conditions at critical gain, and a means of computing K_c for a given system. This is done by translating the magnitude curve vertically until the above conditions are satisfied. The open-loop spectrum shown here describes the transmission once around the loop, and it is from this transmission that the stability of the feedback system can be assessed.

c.   In the following simplified picture, we assume that one cycle of a sinusoidal signal appears at some point in the loop, and is then transmitted around the loop to its point of origin. If the loop gain were somewhat greater than K_{1c} [curve ① in (b)] then at the angular velocity ω_s the transmission would be somewhat greater than unity, and the sinusoidal signal within the loop would continue to grow in amplitude even though the original disturbance had been withdrawn. On the other hand, if $K_1 < K_{1c}$ (as in curve ②) the transmission at ω_s is less than unity and the sinusoid would eventually disappear. The unstable system is distinguished by the fact that it will continue to oscillate even after an initial disturbance has disappeared since the signal can continue to propagate and grow in magnitude around the closed loop. In contrast, the stable system attenuates any such signal, and eventually returns to a quiescent state.

d.   The question of instability as described above would seem to depend upon the initial existence of a sinusoidal signal, because the response was described in those terms. However, this is not the case. All feedback systems are continually beset by disturbances having a large variety of temporal forms, and practically any such signal waveform can be resolved into a frequency spectrum by means of the Fourier transform. Therefore the presence of a sinusoid of angular velocity ω_s is practically assured, simply because the Fourier spectrum of an arbitrary time signal is continuous and contains all frequencies.

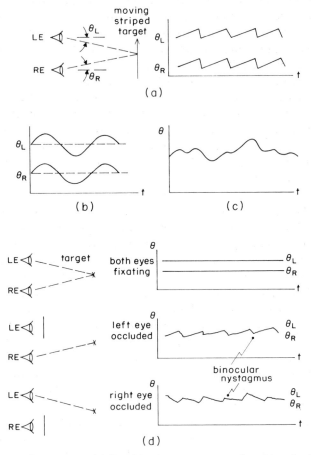

Fig. 7-4. Forms of nystagmus. (a) Optokinetic nystagmus produced by viewing a moving target. (b) Tracking of a target which moves horizontally in a sinusoidal, oscillating manner. (c) Pendular nystagmus in the presence of a fixed target. (d) Latent nystagmus observed in both eyes when viewing a fixed target with one eye occluded.

7   STABILITY

a   *Nystagmus* is defined as an oscillatory or periodic motion of the eyes which may be induced by proper stimuli, or may arise spontaneously from a variety of pathological causes. Although it is only the latter, the self-induced nystagmus, that may be properly classified as unstable, a brief description of the different types may be informative.

b.   A moving target, such as a vertically striped rotating drum, will induce a saw-tooth pattern of eye movement (a), in which periods of eye tracking are interrupted when the eye returns from a limiting position. This is a forced oscillation, termed *optokinetic nystagmus,* and is readily induced by a stimulus such as the drum, or simply by viewing objects from a moving train. It is quite normal, and may be used for diagnostic examination of neurological disorders (Jung and Kornhuber, 1964).

c.   Another example of forced oscillations in eye movement occurs when tracking a target oscillating sinusoidally along a straight line (Dallos and Jones, 1963; Fender and Nye, 1961; St.-Cyr and Fender, 1969; Stark, Vossius and Young, 1962). The sinusoidal behavior of the eye position control system exhibits an interesting adaptive characteristic that requires a careful interpretation of the frequency spectrum.

d   The above examples have been introduced to provide a clear distinction between *forced* and *free* or *self-induced* oscillations. The latter may be evidence for instability as found, for example, in *pendular nystagmus,* and shown in (c). Pendular nystagmus is characterized by a smooth and approximately sinusoidal oscillation that persists for long periods in the absence of any moving stimuli. Such oscillations are observed following fatigue, brain concussion, and even when the eyes are closed. They are also found in eyes with poor vision for which the fixation reflex is poor or absent.

e.   Latent nystagmus, as shown in (d) is not evident in normal binocular vision, but on covering either eye a saw-tooth oscillation appears in both eyes. The quick phase is always toward the fixating eye (Alpern, 1962). The pathology giving rise to latent nystagmus is unknown.

f.   Instability in physiological systems does not always lead to periodic behavior as this example might suggest, but oscillations are not uncommon. They are seldom pure sinusoids and although the relations developed in the previous sections are too elementary for most physiological systems, the basic concepts are sound and may provide some physical insight into the factors conducive to instability.

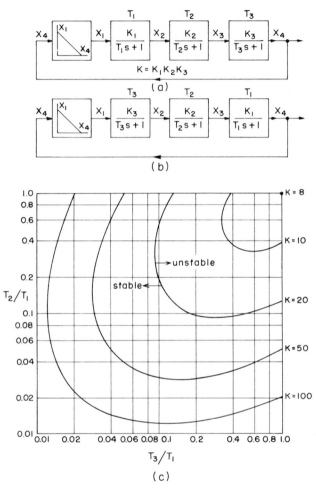

Fig. 7-5. Stability boundaries for a third-order feedback regulator. (a) and (b) are identical noninteracting systems except that the sequence of components has been changed. The stability boundary in (c) is independent of this sequence. This figure is drawn on the basis that T_1 is the largest time constant.

a. Inasmuch as stability is a property of the entire system, it is very difficult to describe the contributions of individual components. However, for linear systems, a few general observations may be made, while for nonlinear systems each must be treated individually.

b. The behavior of the third-order system shown here can be described by means of four parameters; the gain K, and the three time constants, T_1, T_2, T_3. The two block diagrams (a) and (b) show the same system except for a different sequence of the blocks. From the discussion of noninteracting components in Chapter 3, it follows that the sequence of the blocks has no effect upon the behavior. Furthermore, it can be shown that the critical gain K_c depends only upon the time constant ratios T_2/T_1 and T_3/T_1. In the case of interacting components, the relations are not quite as simple as those shown here, and calculation of the stability boundary is slightly more involved.

c. Part (c) of the figure represents the stability boundary surface that separates the stable from the unstable systems. This surface is drawn in *parameter space*, the coordinates of which are the gain and the two time constant ratios given above. The gain is plotted vertically above the paper, and the surface has the form of an inverted cone whose apex lies eight units above the point 1, 1. Any system whose three parameters locate a point lying below the surface is stable, while all points lying above the surface denote unstable systems.

d. For convenience, curves are drawn representing the intersections with the surface of planes at distances of 10, 20, 50, and 100 units above the paper. For example, when $T_2/T_1 = 0.3$, $T_3/T_1 = 0.2$, and $K = 20$, the system is unstable, whereas it would be stable if $K = 10$.

e. Examination of this figure will show that K_c is a minimum when all time constants are equal. Conversely, K_c becomes larger as the several time constants become more unequal. Although this figure was constructed on the basis that T_1 is the largest time constant, it could have been constructed equally well for the case in which T_1 was the smallest. The time-constant ratios would then all be greater than unity.

f. This figure was constructed from an algebraic criterion for stability, but could have been obtained from the log-frequency plots as suggest by Fig. 7-3. The fact that stability depends upon time constant ratios rather than their absolute values is readily seen from such spectra, where a change in all time constants, preserving the ratios, simply shifts the spectrum along the ω-axis.

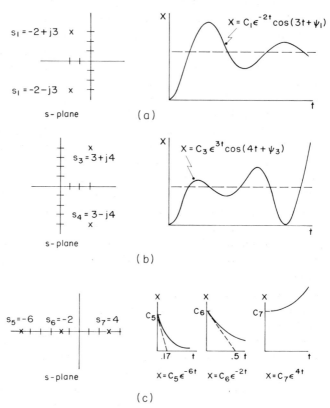

Fig. 7-6. Location of stable and unstable roots on the s-plane. (a) Pair of complex roots with negative real parts and the corresponding damped oscillatory transient. (b) Complex roots with positive real parts and the corresponding oscillatory transient with growing magnitude. (c) Real roots with positive and negative signs.

7 STABILITY

a. The previous discussion regarding the onset of instability in the sinusoidal domain can be paralleled in part by a similar treatment in the time domain. To do this we employ the s-plane and consider the effect of root location on system stability.

b. As a matter of review, (a) shows a pair of complex roots with negative real parts and the corresponding transient behavior. The transient disappears with a time constant equal to the negative of the reciprocal of the real part. The constants appearing in the solution (C_1 and ψ_1) are evaluated to fit the initial conditions, but do not affect the character of the damping.

c. When the roots have positive real parts, as in (b), the transients grow with time, the amplitude increasing exponentially until some saturation effect sets in. Although the responses shown here pertain to a step function disturbance, transients of a similar character would appear for disturbances of any form.

d. In addition to the complex roots described above, real roots are also possible. Three cases are shown in (c), from which it is evident that a positive real root will result in a transient that increases exponentially with time.

e. In technological systems the occurrence of real positive roots is not as likely as finding complex roots with positive real parts. This is because the appearance of a positive real root requires a very specific kind of compartment that, although entirely possible, is not prevalent. It is believed that much the same situation probably holds for physiological systems.

f. From the above examples, one reaches the conclusion that a system is stable only if all roots of the characteristic equation have negative real parts, or, which is entirely equivalent, if all roots lie in the left half of the s-plane. It can be shown that this criterion is equivalent to the one stated previously for the sinusoidal domain (Section 7.4).

g. From the above stability criterion, stated in terms of root location, it is possible to derive other criteria that may be applied directly to the coefficients of the characteristic equation and thus avoid the necessity of actually calculating the roots. Two such algebraic criteria are those given by Hurwitz, and by Routh (Wilts, 1960).

h. Emphasis should be put upon the fact that a single root in the right half plane is sufficient to make the system unstable, whereas stability makes it necessary that all roots lie in the left half plane.

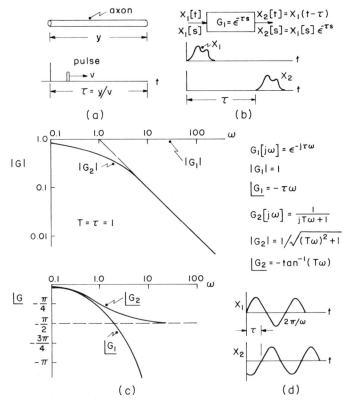

Fig. 7-7. Transport lag (τ) and some of its properties. (a) Neural conduction time (b) Generalized transport lag process with its associated transfer function G_1. (c) Frequency spectrum for transport lag as compared with that for a first-order process (G_2) where $\tau = T$. (d) Transmission of a sinusoidal signal with phase shift but without attenuation.

7 STABILITY

a.   Neural pulses propagate along a nerve fiber at a constant velocity v, the magnitude of which varies with fiber diameter. The pulse is transmitted without change in shape, and a pattern of pulses is propagated without distortion a distance y meters in a time $\tau = y/v$ as shown in (a).

b.   A more general transport lag element, as in (b), transmits a signal of arbitrary shape, X_1, without any modification or distortion, to produce the output signal X_2 of identical form but delayed in time by τ sec. It can be shown that the transfer function for such a component is $\exp(-\tau s)$, or for sinusoidal signals it is $\exp(-j\tau\omega)$. Sinusoids are transmitted without any change in amplitude but with a phase shift of $-\tau\omega$ rad.

c.   The spectrum for an idealized transport lag, G_1, is compared with that for a first-order component, G_2, in (c). The most striking feature is that while the phase of G_2 approaches a maximum value of $-\pi/2$, that for G_1 increases without limit at high frequencies. The comparison has been made arbitrarily for $\tau = T = 1$.

d.   The neural system is almost a perfect example of transport lag in that there is no attenuation nor loss of pulses during transmission. Pulse conduction velocities will range from 1 to 100 m/sec and will therefore exhibit transport lags of from a few milliseconds to about a second. Later discussion will develop the concept that it is not the absolute value of the transport lag that is significant, but rather its value relative to the time constants in the system, and lags of but a few milliseconds may have profound effects.

e.   The circulatory system may also be viewed as a communication channel carrying thermal and hormonal signals. Blood flow is the principal means by which heat is transported throughout the body, and blood temperature may be considered a thermal signal. Hormone concentration in the blood stream is a chemical signal, and the finite circulation time from source to target organ results in a transportation lag. However, with both thermal and hormonal signals, diffusion may well alter the "signal profile," with the consequence that the signal is not transmitted without distortion, in which case the pure transport lag described above is only an approximation to the actual physical process.

f.   The large phase shift associated with transport lag may be better understood if this process is viewed as one of infinite order. That is, the mathematical representation of transport is that of an infinite number of compartments coupled in an interacting manner. The very great number of these compartments results in a phase angle that increases without limit in the manner shown.

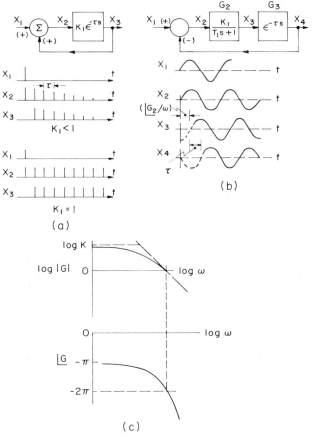

Fig. 7-8. Feedback system with transport lag. (a) Pulse signals in a system with positive feedback and loop gains of less than, and equal to, unity. (b) Sinusoidal signals with loop transmission at the stability limit. The initial portions of X_3 and X_4 (dotted lines) are schematic only in that details of a transient component are omitted. Furthermore, the value of ω must be so chosen as to establish the conditions for critical gain. (c) Frequency spectrum for a system at critical gain (stability limit) showing $|G| = 1$ when $<G = -2\pi$.

a. The introduction of a component with transport lag into a feedback system can have rather surprising consequences. Consider, first, an ideal amplifier of gain K_1 with a time lag $\exp(-\tau s)$ connected to form a positive feedback system as in (a). The amplifier will emit a pulse K_1 times the magnitude of the one received, and this pulse is returned τ sec later to the amplifier input.

b. Let one pulse be introduced at X_1. The resulting sequence of pulses for $K_1 < 1$ is attenuated, and the pulse train disappears. However, if $K_1 = 1$ the pulse train continues indefinitely with the period τ. Any change in K_1 from the value of unity will result in either an attenuation in pulse height as shown, or a continuing increase in pulse height until some component saturates.

c. The case for $K_1 = 1$ might be likened to a pacemaker cell, although it is extremely unlikely that any pacemaker is this simple. This pacemaker, or oscillator, will be stable only if it continues to generate pulses of equal height indefinitely. Any variation in the value of K_1 from unity will be cumulative and there is nothing in this circuit to compensate for such a disturbance.

d. Another feedback system, containing a first-order process together with transport lag, is shown in (b). If a sinusoid is initially introduced at X_1, X_2 will be in phase, but the phase shifts through G_2 and G_3 can conceivably total $-180°$ at that frequency. The signal X_4 is therefore $-180°$ out of phase with X_1, but after passing through the summing point it is in phase with X_2. The initial input signal is therefore no longer needed and the sinusoid once started around the loop will perpetuate itself. This is the case of critical gain, the boundary between stable and unstable operation.

e. Using the data from Fig. 7-7, it is possible to construct the spectrum for the first-order system with transport lag shown in (c). In the absence of transport lag, a first-order system can never become unstable simply because the total phase shift can never exceed $-270°$. On the other hand, a feedback system containing *only* transport lag but with a gain of greater than one will always be unstable. From these considerations one can develop the following statements: (1) The addition of transport lag to a first-order feedback system may well lead to instability. (2) The addition of a first-order process to a feedback system containing only transport lag may stabilize an otherwise unstable system.

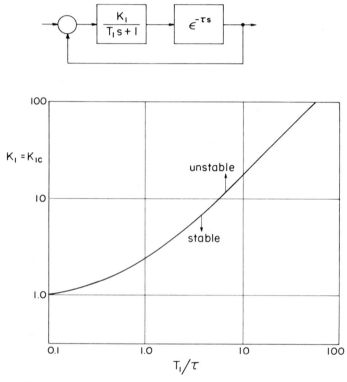

Fig. 7-9. Stability boundary for a feedback system with a first-order process and transport lag. The curve denotes the stability boundary for which $K_1 = K_{1c}$. All points above the curve $(K_1 > K_{1c})$ represent unstable systems. From Jones, "Biological Engineering," H. P. Schwan (ed.), copyright 1969, McGraw-Hill, Inc.; used with permission of McGraw-Hill Book Co.

7 STABILITY

a. Some idea of the effect of transport lag upon the stability of a feedback regulator may be gained from an examination of this simple system, containing only one first-order process in addition to the transport lag. The critical gain is given as a function of the ratio T_1/τ (Jones, 1969).

b. Calculation of the critical gain for a system of this character is perhaps best carried out by means of the logarithmic frequency spectrum in a manner similar to that shown in Fig. 7-8. It will be noted that although the first-order process could never result in an unstable feedback system by itself, in the presence of transport lag instability occurs for rather low values of the gain, K_1.

c. At low values of the ratio T_1/τ, the stability boundary becomes asymptotic to $K_{1c} = 1$. This is equivalent to a feedback system having only transport lag, for which the critical gain is unity. The system will become unstable with gains of greater than one. The presence of the first-order process results in higher values of K_{1c}; the first-order process may be said to stabilize the system.

d. At the other end of the T_1/τ-axis, where τ becomes negligible with respect to T_1, the critical gain increases without limit. That is, the first-order process can never produce an unstable feedback system by itself. It is the presence of the transport lag that results in values of K_{1c} less than infinity.

e. This figure, as well as Fig. 7-5, represent *parametric space* in which some property of the system is given as a function of the parameters. The coordinates are either individual parameters, or functions of the parameters, such as the time constant ratios. In this time lag system, the critical gain is a function of the ratio T_1/τ, whereas in Fig. 7-5 the quantity K_c depended upon two ratios. These figures again show that it is the relative magnitudes of these quantities that is significant, and not their actual values (Choksy, 1962).

f. Although neural conduction times may range from milliseconds to seconds, the time constants associated with muscular response are also of this order of magnitude. One concludes that transport lag may have a significant effect on stability, but this observation must always be examined in more detail for each particular case. A similar statement may be made for circulation time, as it is only in relation to the dynamic behavior of the associated processes that one can evaluate the contribution of this transport lag.

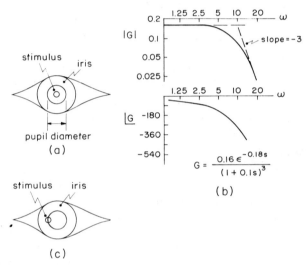

Fig. 7-10. Experimental study of the pupil reflex. (a) Stimulus is a collimated beam of light having a diameter less than the minimum pupil diameter. (b) Frequency spectrum and transfer function from the open-loop experiment in (a). The light beam intensity is modulated sinusoidally and the pupil diameter is measured photoelectrically. (c) Stimulus is a beam of constant intensity located at the iris margin. (Stark and Sherman, 1957; Stark, 1959).

a. The desirability of measuring the open-loop characteristics of a homeostatic system will be apparent from previous discussions. At the same time, many problems arise in the development of such experiments; some of these are discussed in Sections 5.15, 7.12, and 8.11.

b. One of the first measurements of open-loop transmission was made with the pupillary reflex arc (Stark and Sherman, 1957; Stark, 1959). This feedback system, because of its peculiar nature, can be interrupted externally, as indicated in (a). If the stimulating light beam is made smaller than the smallest pupil opening, then there is no feedback (the iris does not affect the stimulus) and measurement of excursions in pupil diameter as the stimulus intensity is modulated yield a true value of open-loop transmission.

c. The results of such an experiment for sinusoidal stimuli are shown by the spectrum in (b). The low frequency asymptote is approximately 0.16, a value for loop gain that is very small compared to the examples described earlier. The high frequency asymptote has a slope of -3, and since only one corner frequency of 10 rad/sec is apparent, the process is interpreted as having three compartments, each with a time constant of 0.1 sec. However, the phase angle exceeds $270°$ at high frequencies, which suggests a transport lag of approximately 0.18 sec (see Section 7.8).

d. One test of these measurements lies in the possibility of inducing self-oscillations in the reflex arc by suitably increasing the loop gain to its critical value in the manner described by Fig. 7.3. An increase in loop gain can be accomplished by locating the stimulating light beam at the iris margin as shown in (c). By proper choice of beam area and location it is possible to achieve critical gain. Under constant stimulus conditions (beam not modulated externally), it has been found that the reflex arc will break into continuing oscillation (Stark and Cornsweet, 1958; Stark, 1959). Even though the reflex contains nonlinearities, the actual frequency of oscillation was close to that predicted by linear analysis from the frequency spectrum. That is, the oscillating frequency closely matched that at which the magnitude of the spectrum was unity, and its phase $-180°$.

e. Careful analysis of the experiment suggested by the figure in (c) reveals that the reflex arc has a gain that varies with the pupil diameter so that application of linear techniques to assess stability of this nonlinear system is questionable. Despite this fact, it is of interest to note that linear approximations can be of use.

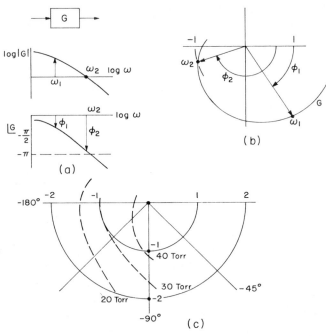

Fig. 7-11. Polar plots of the frequency spectrum and their use in predicting the onset of instability. (a) Conventional logarithmic plot for a system whose open-loop transfer function is G. (b) A polar plot of this spectrum in which a phasor is plotted for each value of ω having the magnitude and phase as given in (a). The magnitude plotted is the actual value, not its logarithm. (c) Experimentally obtained polar plots for the blood pressure regulating system in dogs for three values of mean cerebral perfusion pressure (MCPP). From K. Sagawa, O. Taylor, and A. C. Guyton, *Amer. J. Physiol.* **201,** 1961, p. 1164, by permission. The curve for 20 mm Hg is for a different animal than the other two curves. With high MCPP, the polar plot lies inside the point -1, whereas with lower values of MCPP, the curve passes through -1, or surrounds it. In the latter case, the closed-loop regulator would be deemed unstable.

a. The frequency spectrum may also be plotted in polar coordinates, as in this figure. Heretofore, the process (G) has been represented by its spectrum drawn in logarithmic coordinates as in (a), with the magnitude and angle portions separated.

b. The two parts of the spectrum may be combined on a polar plot (b), where the transmission at each ω serves to define a phasor having the magnitude and angle of the sinusoidal transmission. However, the polar plot uses the actual magnitude of G, *not* its logarithm. Phasors are drawn for as many values of ω as are desired, and their tips connected by a smooth curve. Such a polar plot is also termed a *Nyquist diagram*.

c. The question of stability then turns upon the behavior of the polar plot in the vicinity of the point -1, because this point corresponds to the condition that the magnitude of G is unity, and its phase angle is $-180°$. This angle, together with the $180°$ contributed by the sign inversion, provides the $360°$ necessary for continued oscillations.

d. It has been observed that under certain conditions of hemorrhage in dogs, continuing oscillations in blood pressure would appear, the frequency of which was unrelated to respiration or cardiac rate. These are termed *Mayer waves,* and have been postulated to be evidence of instability in the blood pressure regulating mechanism. Evidence supporting this hypothesis is presented below (Cherniack *et al.,* 1969).

e. In cross circulation experiments, it has been possible to control the mean cerebral perfusion pressure (MCPP) while subjecting it to sinusoidal variations (Sagawa *et al.,* 1961). Typical results are plotted in (c), where it appears that at mean pressures below about 30 Torr, the point -1 will be enclosed, and the closed-loop regulator would become unstable.

f. Using servo control of the cerebral perfusion pressure so as to arbitrarily set the mean pressure without effecting the oscillations, it has been demonstrated that the pressure regulating system will indeed break into maintained oscillations when the mean pressure is reduced (Sagawa *et al.,* 1962). The period of these oscillations "coincided almost exactly" with that predicted; namely, the frequency (ω) on the polar plot at which it passed through the point -1.

g. Further confirmation of this hypothesis was obtained with animals following controlled hemorrhage (Race and Rosenbaum, 1966). Upon loss of a fixed amount of blood, the introduction of a disturbance caused the pressure to exhibit a transient oscillation whose frequency increased and damping decreased with increasing blood loss.

In view of the previous discussion, it would be quite appropriate to ask why we do not see more evidence of physiological instability. There are a number of aspects to this question, but first some further examples of oscillatory behavior will be cited.

(1)   *Hippus* is an oscillation in pupil size that is observed under conditions of constant general illumination (not the collimated beam of Fig. 7-10), and may have a frequency as high as 2 Hz. Although there is considerable disagreement in the literature as to the etiology and significance of these oscillations, some general observations will be made. Lowenstein and Lowenfeld (1962) report that oscillations are found in most normal subjects when exposed to long-lasting light stimulation, but that the oscillations are reduced or absent if the afferent or efferent pathways of the light reflex are impaired. On the other hand, many writers interpret hippus as pathological, and ascribe this behavior to a variety of lesions within the central nervous system. A reasonable appraisal of these viewpoints would suggest that the reflex is normally close to a point of instability represented by transient oscillations having but little damping, so that only minor alterations in the transmission of the several pathways are required to induce a continuing oscillatory behavior. This is reminescent of a similar situation in neuromuscular control (Section 5.14).

(2)   *Oscillations in erythropoiesis* have been reported in the dog (Morley and Stohlman, 1969). Erythrocyte concentration is regulated by a feedback system that includes a sensor of oxyhemoglobin (probably in the kidneys) and the production of hemoglobin in the bone marrow. This production rate is under the direct control of the erythrocyte concentration. A salient feature of this regulator is the fact that the red cells have a life span in the circulatory system of approximately 120 days. The hemoglobin compartment thus has an input which is the production rate of red cells in the bone marrow, but the compartment has a loss function that does not depend upon the concentration of red cells, but upon their age. It may be shown that this loss of senescent cells is equivalent to a time lag in the feedback system of about 120 days, and that this time lag may have significant effects on the system stability (see Section 7.10). The existence of oscillations in a feedback system with time lag is strongly suggestive of an unstable regime, but further evidence is needed to substantiate such an hypothesis.

(3)   Although limb movement is normally well damped, one does find examples in which oscillations in limb position persist for extended periods, and in experimental animals it is possible to demonstrate main-

7   STABILITY

tained oscillations (*clonus*). Granit (1959) has stated, "When considering the circuit of the stretch reflex schematically, it is perhaps less surprising that it oscillates in clonus than that it does not always do so when recording is isotonic." Clonus is readily observed in decerebrate cats with isotonic recording, but not under isometric conditions. The spindles as well as recurrent inhibition play a significant part in clonic activity, but the Golgi tendon organs have little effect. It should not be concluded that all tremors observed in limbs are attributable to instability in the stretch reflex, but rather that the latter can become unstable under proper conditions. Neuromuscular oscillations also appear in shivering as a co-contraction of antagonistic muscles. It is controlled centrally, probably by hypothalamic structures, but the shivering frequency seems to be determined peripherally. Furthermore, shivering requires intact afferent and efferent pathways to the muscle, as well as the cerebellum. It is probably too simple to suggest that shivering is a manifestation of instability in the stretch reflexes, but the fact that these reflexes are lightly damped might well be conducive to the continued oscillations observed in shivering.

(4) *Cheyne–Stokes breathing* may be described as a slow oscillation in the breathing pattern superimposed upon the normal respiratory frequency. The tidal volume increases from zero to a maximum value and then subsides until breathing ceases, and after an interval this pattern is resumed. Although this oscillatory behavior is ascribed to a variety of causes, it can be induced in normal subjects by hyperventilation. Computer studies suggest that the following factors are significant (Horgan and Lange, 1962): (a) hyperventilation, (b) decreasing lung volume, (c) decreasing blood pH, and (d) increasing circulation time. Although Cheyne–Stokes breathing is described above as oscillatory, it is far from sinusoidal in character, and thus hardly amenable to linear analysis.

These and other examples provide fairly convincing evidence that homeostatic systems may indeed become unstable, and it is possible that instability, or the approach thereto, is more widespread than is ordinarily realized. Goodman (1964), in an experimental study of ventilatory rate in man, obtained evidence suggesting the existence of some five frequencies having periods ranging from 1–50 min in a normal subject. These are not subharmonics of the breathing frequency, but appear to be independent and almost periodic oscillations. They might be interpreted as evidence for instability in a number of feedback loops associated with the respiratory system (Goodman *et al.*, 1966).

The maintenance of oscillations in all of the above examples is assumed to rest upon two factors. The first is an unstable operating point, which leads to oscillations of increasing amplitude. Second, some form of saturation serves to limit the amplitude of oscillation. These two factors are sufficient to account for the continuing oscillations, and are not too unrealistic.

In addition to oscillatory behavior as evidence for instability, it was shown in Section 7.7 that the presence of a real positive root of the characteristic equation could also lead to an unbounded response, although some form of saturation would ultimately restrict the variables to finite but probably abnormal values. In the case of such disorders as glaucoma, hypertension, diabetes, and hyperthyroidism, oscillations are not normally observed but the system variables leave their normal operating regions and take on abnormal values. Should this behavior be termed unstable? While this is a possible explanation, it is also conceivable that some parameter has changed such as to move the operating point outside of the region of normal regulation. In the case of the fluid flow examples treated earlier, an increase in an outflow resistance can cause a tank to overflow and render the feedback mechanism inoperable. Although the system is unusable, it would not be termed unstable according to the previous definitions. The observation of a final state, either one of oscillation or of saturation, does not in itself reveal how the system got there. We can only infer that instability is a possible, but not the only, cause.

The basic concept of stability is associated with the notion that a bounded disturbance to a stable system will produce a bounded response. For "bounded," one may substitute finite, or possibly small, and then state that small disturbances to a stable system can produce only small responses. If the system is stable, then the response will remain small for all small disturbances, but on the other hand any small disturbance to an unstable system will initiate an unbounded response. The unstable system exhibits an irreversible change in its behavior following a disturbance, and in many cases this new behavior causes the sytem variables to take on values beyond their physical capability. An unstable system may be disrupted by any disturbance that appears, and is no longer able to function in its normal manner.

Another facet to the concept of stability has to do with what I shall term its subjective aspects. This means that the definition of stability cannot be assumed independently of the system being considered and its normal functions. An example or two will elucidate. The body thermostat presumeably functions to keep the internal temperature constant, and a continuing oscillation of the internal temperature might well be considered unstable and pathological. On the other hand, rhythms such as the cardiac or respiratory cycles are normal, and instability must be defined in terms of upsets to the normal cyclic behavior. This difference in viewpoint in the application of the stability concept has nothing to do with any differences between technological and biological systems, but rather between systems having a constant and a cyclic steady state.

In addition to the two cases mentioned in the previous paragraph, it will be found that a number of different definitions of stability are required to provide for a number of physical situations. Among these definitions the following may prove of interest (Cunningham, 1958). A system is said to be *asymptotically stable* if, following a disturbance, the system returns asymptotically to its previous steady state, or if the disturbance is prolonged, to a new steady state. This is the case discussed in the previous sections. An *absolutely stable* system is one that remains stable for all values of its parameters. Although this definition includes some trivial cases, it also includes high-order systems having transfer functions of the proper kind. A system is said to be *conditionally stable* if it is stable for both high and low values of loop gain, but is unstable for intermediate values of gain. Technologically, such systems are acceptable providing ways can be found to keep them from operating in this intermediate range. *Orbital stability* refers to the ability of an oscillating or periodic system to maintain its frequency and amplitude.

The above definitions have been borrowed from the mathematical theory of stability, but slightly reworded in some instances. Another term is *global stability,* used to describe a system that remains stable for all possible normal operating conditions; it is thus a more restrictive category than that of absolutely stable systems. Physiological regulators must obviously exhibit global stability in that they are stable under all *normal* operating conditions, but they need not be absolutely stable in that pathological conditions may lead to instability.

Stability is a property of the entire system, its interconnections, and the magnitude of all its parameters, and is not affected by the disturbance imposed by the environment. However, the fact that it takes a disturbance to initiate unstable behavior, if such exists, may appear to be a paradox. Conceptually we can visualize any physical system as operating in a quiescent steady state, and the question is then one of assessing the stability of that operating point. The possible operating points of a physical system are of two varieties; that is, either stable or unstable, and the stability criteria enable one to distinguish between the two. In the unstable case, any disturbance will cause the sytem to move away from that steady state and never return. All physical systems are continually beset by small disturbances that affect all the physical variables if feedback is present, and may have a wide variety of temporal forms. Any such disturbance is sufficient to "test" the system for stability, and it is not necessary to impose a test disturbance for this purpose. Disturbances frequently alter the operating point so it becomes necessary to test all possible operating points for stability. For example, a change in loop gain will also change the operating point, but we have already seen that as the gain is increased, most systems become unstable. Similarly, changes in the other system parameters will usually alter the operating point and may affect stability.

If the loop gain is adjusted to be at the critical value K_c, a transient response induced by any disturbance will be a continuing oscillation of constant amplitude, as was implied by Fig. 7-2c. However, it is pertinent to enquire as to the practical significance of this conclusion. Continuing oscillation at a constant amplitude requires that the two relations, $|G| = 1$, and $\angle G = -360°$, be precisely satisfied, and the probability of this occurring for any finite period of time is vanishingly small. This conclusion follows from the knowledge that the ever-present disturbances will prevent these relations from being satisfied over any reasonable interval. Thus, while operation at the stability boundary is conceptually possible, it is not a stable operating point, and could not be used to form a stable oscillator. The physical system is clearly either stable of unstable. Further discussion of this point will be found on p. 315.

The previous discussion in this chapter has more or less implied that the onset of instability occurs as a consequence of an increase in loop gain. Although this is to some extent an engineering viewpoint, we should like to point out some physiological implications. Furthermore, a change in any of the system parameters may lead to instability, if that change brings the system closer to the stability boundary. With low loop gains, most systems are stable, and it is when the gain exceeds K_c that instability ensues. On the other hand, sensitivity to disturbing factors is reduced by an increase in gain (Section 4.9) so that the demands of sensitivity and stability conflict when it comes to selecting gain. In engineering practice, the normal procedure would be to select a gain to meet the sensitivity requirements. If this gain resulted in an unstable system, additional stabilizing components and pathways would be added that would have no effect upon the gain required for the sensitivity reduction, but on the other hand would provide the damping necessary for stable operation. Has evolution acted in a somewhat similar manner for homeostatic systems? There is no body of evidence bearing on this question beyond the mere fact that biological processes are normally stable, and they do exhibit low sensitivity. An interesting conjecture would suggest that evolutionary processes have acted to reduce the sensitivity of feedback systems. If this has occurred by increasing loop gain, then one should find that homeostats are working somewhere near the stability limit. However, it may well be that evolution has resulted in a decreased sensitivity by some other means than an increase in gain, or that suitable

7 STABILITY

dynamic compensating processes have been developed to permit operation at high gain. Information regarding this point is meagre, although Adolph (1968) does discuss some of the pertinent aspects.

Motivation for a concern over stability is quite different in engineering systems from that generated by a study of organisms. In engineering design, stability is a prime concern, and in most instances components must be added to the basic system to assure stability. In contrast, homeostatic processes available for study are normally stable except for pathological cases, as has been suggested. Unstable processes that may have evolved along the line are not possible subjects for our analysis. But there is more motivation for the enquiring mind than the above obvious distinction. The mere complexity of homeostatic systems raises the basic question of how stability is in fact achieved, and until that question has been at least partially answered we cannot feel secure that our understanding of those systems is in any sense adequate. It should also be borne in mind that stability as we have discussed it refers to a model of the prototype system, and the conclusions are only as valid as the model. Stability becomes one measure of the competence of the model, in that if both model and prototype can be thrown into instability by the same parametric changes, there is reason for having greater confidence in the mathematical representation.

Epitome 7.14

1. Although common usage associates the term stability with a constancy of behavior, when applied to a regulating system stability is more likely to connote an avoidance of disruptive action. In any case, stability can mean many things and so must be carefully defined in the context in which is is applied.

2. A regulating system of the third order or higher will always exhibit transient oscillations when disturbed if the loop gain is sufficiently high. Moreover, these oscillations will not be damped but may increase in amplitude without limit if the gain is increased.

3. The value of the loop gain at the boundary between transient oscillations that decrease with time and those that increase with time is termed the critical gain. At critical gain, any transient oscillations would in theory continue at a constant amplitude.

4. A feedback system having critical gain would exhibit a frequency spectrum, the amplitude of which would be unity at the angular velocity for which the total phase shift was 360°. In such a system a sinusoidal

signal, introduced at any point, would propogate once around the loop without any change in amplitude or phase, and could thus perpetuate itself.

5. As an example of biological instability, one may cite nystagmus in which periodic eye motion persists even with stationary targets. However, one must carefully distinguish between such periodic motion and the motion induced by targets having continuous or periodic motion; these latter stimuli produced a forced oscillation in the eye position, and are not a consequence of instability.

6. In a feedback system consisting of first-order processes, the critical gain depends only upon the time constant ratios, and not their absolute values. Stability boundaries may be drawn in parametric space so as to define regions in which the system is stable or unstable.

7. Roots of the characteristic equation having positive real parts lead to transients that grow (rather than decrease) with time. A system is stable only if all roots lie in the left half of the s-plane.

8. Transport lag is defined as a process in which a signal is transmitted without any change in shape, but with a fixed delay. The frequency spectrum for such a process exhibits no change in magnitude with angular velocity, but a continually increasing phase angle.

9. A feedback system having only transport lag is unstable if the loop gain is unity or greater. The stability of a system having both a first-order process and transport lag may be evaluated from the frequency spectrum in the manner previously described.

10. The stability boundary for a feedback system containing one first-order process and a transport lag shows that the critical gain is only slightly greater than one for T_1/τ ratios of unity or less. However, K_c becomes high as T_1/τ approaches large values.

11. The pupillary reflex is unique in that open-loop tests can be made without surgical intervention. Furthermore, since gain may be independently adjusted, it is possible to obtain an unstable feedback system, the performance of which is very close to that predicted from the open-loop spectrum.

12. The blood pressure regulator will become unstable if the mean arterial pressure is reduced. This reduction apparently has the effect of increasing loop gain, and the point of instability can be quite well predicted from the open-loop frequency spectrum.

13. Although evidence for instability is found in a variety of physiological systems, the significant question is how stability is in fact achieved in all normal homeostats. Since the criteria for stability must be applied to the mathematical models, such tests may not contribute greatly to the above question, but rather provide a means of examining and improving model validity.

1. For the feedback system in Fig. 7-1, let $\dot{Q}_1 = \dot{Q}_{10} - K_3 Y_3$, where $\dot{Q}_{10}$ is a constant.
   (a) Show that loop gain is $K_1 = K_3 R_3 / \delta$.
   (b) Show that K_1 is dimensionless.

2. Confirm the value of K_c for $T_2/T_1 = 0.1$ and $T_3/T_1 = 0.19$ as given by Fig. 7-5. (Hint: Use a log frequency spectrum.)

3. A second-order feedback system for which $T_1 = 3$ sec and $T_2 = 1$ sec and $K = 10$ is of course stable. If a third compartment is added (T_3), for what value of T_3 will this system become unstable?

4. (a) For a fourth-order system for which $T_1 = T_2 = 1$ and $T_3 = T_4 = 0.5$, find the critical gain K_c.
   (b) If $T_4 = 0$, what will K_c become?

5. (a) Compute and plot the curve for $\angle G_1$ in Fig. 7-7.
   (b) Calculate K_{1c} for the system in Fig. 7-9 for $T_1/\tau = 0.1$ and for 5. Compare your results with the curve in that figure.

6. Repeat Problem 3, but instead of adding a third compartment, add a time lag process, τ. What value of τ will make the system unstable?

7. Figure 7-5 suggests that a third-order system for which $T_2/T_1 = T_3/T_1 = 1.0$ has a critical gain of eight. Confirm this conclusion by any method.

8. From the discussions in this chapter estimate the critical gain for each of the following cases and give the source of your estimate.
   (a) $T_1 = 1$, $T_2 = 3$, $T_3 = 5$;
   (b) $T_1 = 4$, $T_2 = 16$;
   (c) $T_1 = 1$, $\tau = 0.1$;
   (d) $\tau = 0.01$.

9. Refer to the system sketched in Fig. 7-8a.
   (a) Do the decaying pulses have an exponential envelop? Why?
   (b) Sketch a system similar to that in this figure except with a negative feedback. Sketch the response for a single input pulse under both stable and unstable conditions.

10. A process is said to be of *zero order* if the outflow rate is a constant, and therefore independent of the system variables. In terms of fluid flow models, a zero-order process might be a pump that removes fluid at a constant, but possibly adjustable, rate. In a biological process, the active transport of molecules at a rate independent of the concentrations would also be a zero-order process.

(a)   Show that in general, a stable steady state is impossible if an irreversible reaction precedes a zero-order reaction.

(b)   Is this also the case if a reversible process precedes one of zero order?

11.   Estimate the critical gain for the pupil reflex arc from Fig. 7-10 and compare it with the value found experimentally (Stark and Cornsweet, 1958).

References

Adolph, E. F. (1968). "Origins of Physiological Regulations." Academic Press, New York.
Alpern, M. (1962). Movements of the eyes, in "The Eye" (H. Davson, ed.), Vol. 3, p. 86. Academic Press, New York.
Bellman, R. (1953). "Stability Theory of Differential Equations." McGraw-Hill, New York.
Cherniack, N. S., Edelman, N. H., and Fishman, A. P. (1969). Pattern of discharge of respiratory neurons during systemic vasomotor waves. *Amer. J. Physiol.* **217**, 1375–1383.
Choksy, N. H. (1962). Time lag systems. *Progr. Control Eng.* **1**, 17–38.
Cunningham, W. J. (1958). "Introduction to Nonlinear Analysis." McGraw-Hill, New York.
Dallos, P. J., and Jones, R. W. (1963). Learning behavior of the eye fixation control system. *IEEE Trans. Auto. Cont.* **AC-8**, 218–227.
Fender, D. H., and Nye, P. W. (1961). Investigation of the mechanisms of eye movement control. *Kybernetik* **1**, 81–88.
Goodman, L. (1964). Oscillatory behavior of ventilation in resting man. *IEEE Trans. Bio-Med. Eng.* **BME-11**, 82–93.
Goodman, L., Alexander, D. M., and Fleming, D. G. (1966). Oscillatory behavior of respiratory gas exchange in resting man. *IEEE Trans. Bio-Med. Eng.* **BME-13**, 57–64.
Granit, R. (1959). Observations on clonus in the cat's soleus muscle. *An. Fac. Med. Montevideo* **44**, 305–310.
Horgan, J. D., and Lange, R. L. (1962). Analog computer studies of periodic breathing. *IRE Trans. Bio-Med. Electron.* **BME-9**, 221–228.
Jones, R. W., (1969). Biological control mechanisms, in "Biological Engineering." McGraw-Hill, New York.
Jung, R., and Kornhuber, H. H. (1964). Results of electronystagmography in man; the value of optokinetic, vestibular, and spontaneous nystagmus for neurologic diagnosis and research, in "The Oculomotor System" (M. B. Bender, ed.). Harper (Hoeber), New York.
Lowenstein, O., and Loewenfeld, I. E. (1962). The pupil, in "The Eye" (H. Davson, ed.), Vol. 3, pp. 231–267. Academic Press, New York.
Morley, A., and Stohlman, F. Jr. (1969). Erythropoiesis in the dog; the periodic nature of the steady state. *Science* **165**, 1225–1227.
Race, D., and Rosenbaum, M. (1966). Nonrespiratory oscillations in systemic arterial pressure of dogs. *Circ. Res.* **18**, 525–533.
St.-Cyr, C. J., and Fender, D. H. (1969). Nonlinearities of the human oculomotor system: gain. *Vision Res.* **9**, 1235–1246.
Sagawa, K., Taylor, A. E., and Guyton, A. C. (1961). Dynamic performance and stability of cerebral ischemic pressor response. *Amer. J. Physiol.* **201**, 1164–1172.

Sagawa, K., Carrier, O., and Guyton, A. C. (1962). Elicitation of theoretically predicted feedback oscillation in arterial pressure. *Amer. J. Physiol.* **203**, 141–146.

Stark, L. (1959). Stability, oscillations, and noise in the human pupil servomechanism. *Proc. IRE*, **47**, 1925–1939.

Stark, L., and Cornsweet, T. N. (1958). Testing a servoanalytic hypothesis for pupil oscillations. *Science* **127**, 588.

Stark, L., and Sherman, P. M. (1957). A servoanalytic study of consensual pupil reflex to light. *J. Neurophysiol.* **20**, 17–26.

Stark, L., Vossius, G., and Young, L. R. (1962). Predictive control of eye tracking movements. *IRE Trans. Human Factors in Electron.* **HFE-3**, 52–57.

Wilts, C. H. (1960). "Principles of Feedback Control." Addison-Wesley, Reading, Massachusetts.

Distinctive Features of Homeostatic Systems

Chapter 8

Introduction   8.1

In the previous chapters, we have pointed out certain differences that seem to distinguish biological regulators from their technological counterparts. Although these two classes of systems are related by the common bond of negative feedback, there are significant distinctions, not only in their organization but also in the way systems in the two categories are studied and analyzed.

In the first place, the engineer designs the technological regulator with some specific objective in mind, and makes decisions regarding the components that are to compose the system. In contrast, the homeostat is a functioning system that has acquired its precision of operation and complexity over a very long evolutionary period, and whose objective (or objectives) may not be known precisely. Components now a part of the system may no longer best serve their original function, and the identification of those components that are necessary parts of the system poses many problems. It may be more to the point, however, to note that man is not now designing his own homeostats, but rather seeks a fuller understanding of both their normal functioning as well as their pathology.

In the study of system behavior we equate understanding with a knowledge of how the components contribute to total system performance, which includes both the characteristics of each component as well as the manner in which the components are interconnected. A study of this character is a very large order for even the simplest biological process, but nonetheless a necessary program if the science of medicine is to achieve further advances in diagnosis and therapy. To the above program must be added a consideration of the manners in which the many homeostats in one organism interact with each other.

For the most part, technological regulators are very elementary systems compared to even the simplest homeostat. It may well be that, entirely apart from questions of reliability, some of the most essential features of homeostatic behavior derive from this complexity. In this situation, although technological regulators may suggest many questions of function and organization, the investigator is forced to the physiological prototype for his answers.

This chapter attempts to isolate in more detail some aspects of homeostats which seem to distinguish them from technological feedback systems. These features include not only the components themselves and their organization, but also the experimental problems involved in their study.

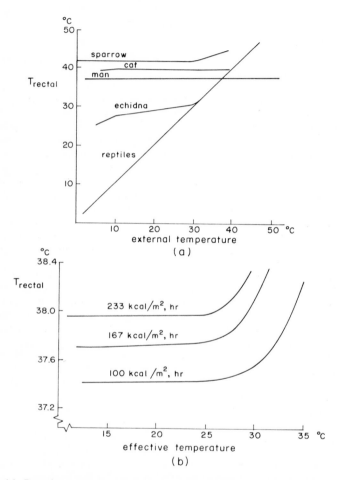

Fig. 8-1. (a) Rectal temperatures of some representative poikilotherms and homeotherms as functions of the external temperature. Similar curves for other species may be found in Precht, Christophersen, and Hensel, 1955. (b) Rectal temperature in man for different external temperatures and work rates (Lind, 1963).

a. The identification of a homeostat may be divided into two parts; first, there is the circumstantial evidence that may be obtained from external observation and appropriate disturbing tests. Second, to complete the identification it is necessary to trace the pathways and components through which the feedback operates.

b. Within the various animal species, one finds organisms whose internal temperatures closely follow that of the surroundings; these are the *poikilotherms,* or cold-blooded animals. On the other hand, most of the higher animals exhibit an internal temperature that is largely independent of the ambient air; these are the *homeotherms.* In between these two extremes are many species that appear to exhibit some regulation of internal temperature over limited ranges. Part (a) of the figure shows some examples; the external temperature is presumeably some average temperature of the surrounding air.

c. While part (a) illustrates the relative independence of rectal temperature from that of the surroundings, part (b) shows a similar independence for heat stresses imposed internally by a work load. Note both the difference in temperature scales in the two figures, as well as the use of an "effective" temperature in (b). The latter is a scale developed to describe subjective sensation, and includes the effects of wind velocity and humidity. Although the subjective aspects of temperature probably have no effect upon the regulating process, this figure does show an essentially constant rectal temperature over a range of effective temperatures as well as the fact that this constant rectal temperature does change slightly with the work load. The latter effect might be interpreted as an increase in temperature gradient required to dissipate the additional losses, or as an actual change in the operating point. The curves in these two figures are difficult to compare because of the difference in scales used.

d. A second kind of evidence for the existence of negative feedback is the location of a sign inversion process. In the case of sweating, it is obvious that the evaporation of moisture serves to enhance the loss of heat from the body and thus decrease the body temperature. The sign inversion will usually be found to be an inherent property of some process within the sytem.

e. The identification can only be completed by locating the signal pathways which serve to couple the several variables in a cyclic manner. The mere fact that some physical quantity remains constant in the face of disturbing factors is at best only circumstantial evidence.

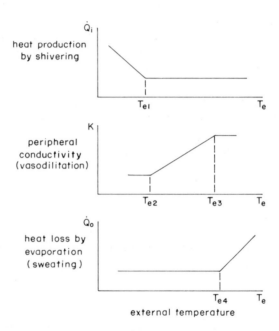

Fig. 8-2. Operating ranges of the three principal temperature regulating mechanisms in man. The four thresholds T_{e1}, T_{e2}, T_{e3}, and T_{e4} are probably independent, but must obviously have such values as to provide continuous regulation over the whole temperature range. The thresholds cannot be given unique numerical values because they depend to some extent upon the pattern of thermal loading and stress.

a. A cursory examination reveals that temperature is regulated not by one mechanism, but by several, each having a range of temperatures over which it is most effective, as illustrated schematically in the figure. The principal mechanisms, shivering, vasomotor control, and sweating, are believed to be independent regulating systems in that each operates over a range of temperatures in which the other mechanisms are at their minimum or maximum effectiveness (see Problem 2).

b. Vasomotor control, operating over the normal, intermediate temperature range, serves to dilate or constrict the peripheral vessels so as to provide a fine control of the heat losses from the skin that is adequate for moderate thermal loads. At T_{e2} the constriction is a maximum, and regulation over lower temperatures must invoke some form of thermogenesis, usually heat production by shivering. At T_{e3} vasodilitation is a maximum and further heat loss requires the onset of sweating.

c. Although this figure suggests the close relationship between these three homeostats it is possible, both experimentally and theoretically, to study each individually on the basis that the other mechanisms are either saturated or cut off. Such an isolation of each feedback mechanism is almost a necessity, at least in a preliminary study, and we shall adopt this viewpoint even though it may be difficult to carry out.

d. There are additional factors that make the isolation of a single homeostat a difficult process. In addition to the three mechanisms mentioned above, there are other processes that aid in temperature regulation in man and other species, such as hormonal control of metabolism, panting, and behavior patterns. These factors can be controlled in laboratory studies, but in the absence of such controlled experiments the interpretation of such observations is extremely difficult.

e. So far, we have mentioned only those processes having an immediate bearing upon temperature regulation, but it must be noted that other homeostats also may be intimately involved. Chief of these are the cardiovascular system and respiration, each of which has a direct connection with thermogenesis, heat transport, or heat loss. One quickly concludes that temperature regulation, and without doubt most of the other homeostats, are coupled to each other in a variety of ways, which makes their individual study very difficult.

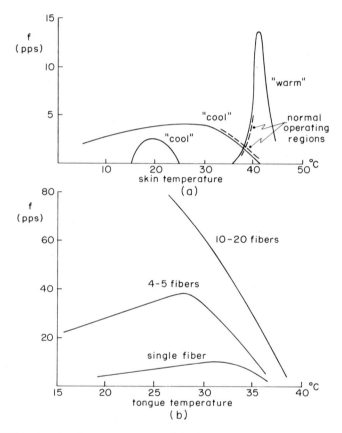

Fig. 8-3. Neural pulse frequencies in fibers from temperature receptors in the cat. (a) Single fibers in the skin (Hensel, Iggo, and Witt, 1960). (b) Fiber bundles in the lingual nerve (Hensel and Zotterman, 1951).

a. Homeostatic systems usually contain neural receptor units, the function of which is to "measure" a specified variable and provide a neural signal that can be used for control and regulation purposes. This section describes the temperature receptors and some of their distinctive properties (Hensel, 1963).

b. Specific temperature receptors are identified as such by virtue of their high sensitivity to temperature change, as well as a high threshold for mechanical stimulation. These are *tonic* units in which the pulse frequency is maintained for long periods at a value proportional to the temperature stimulus.

c. The "cool" receptors are distinguished from the "warm" units by the fact that the temperature for maximum discharge frequency lies below the normal temperature, whereas with the warm receptors the maximum frequency lies above. The slope of the steady-state characteristic in the operating region is thus positive for the warm and negative for the cool receptors. This negative slope probably provides the sign inversion for negative feedback when shivering is the effective mechanism.

d. The characteristics of thermal receptors shown in (a) are double-valued functions of temperature in that there are two possible temperatures having the same firing frequency. Presumeably, the nervous system has ways of assuring operation in the appropriate regions. Part (b) suggests that the distribution of characteristics for the individual receptors is such that the thermal signal from a number of receptors has a range sufficiently broad to encompass all operating conditions.

e. The identification problem for thermal receptors is especially difficult. While the receptors examined for this figure were from the cat, other experiments reveal almost identical properties in fibers from the radial nerve in man (Hensel and Boman, 1960). There is little question that receptors do exist in the skin (as well as at a variety of loci within the body) having these properties, but there is no experimental demonstration as yet to establish these same receptors as the ones responsible for temperature regulation. Furthermore, there is some reason for questioning the assumption that the same receptors serve both sensory and control functions (Jones, 1969).

f. Further studies to identify the receptor structures involved in regulation are even more difficult. In any case there is considerable doubt as to whether the Kraus end-bulbs and Ruffini endings are the thermoreceptors even though this identification is frequently made.

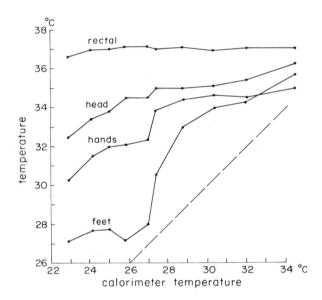

Fig. 8-4. Measured temperature of selected points throughout the body. Note marked change in the temperature distribution as the surrounding (calorimeter) temperature is varied (Du Bois, 1941).

a. In the steady state, the heat losses from the body must equal the heat generated internally plus that absorbed from the environment. However, this relation imposes no requirement that the body temperature be uniform, and as a matter of fact the temperatures throughout the body may be far from uniform, and may even be changing with time, while still maintaining a steady-state energy balance.

b. Temperature over the body surface may range over several degrees as indicated by the figure, and internal temperatures may vary by several tenths of a degree, with the liver frequently being the hottest spot. One may look upon this distribution of temperatures as that required to produce the necessary heat flow to the surface to maintain the energy balance. Regulatory actions of the body thermostat serve to change this pattern of heat flow, and therefore may alter the internal temperature distribution.

c. Clinically, oral or rectal temperature is usually taken to be indicative of the state of the body, but neither of these two quantities is believed to play a significant role in temperature regulation. On the other hand, there is considerable evidence that two "centers," the anterior and posterior nuclei in the hypothalamus, do play essential roles in temperature regulation. An animal loses its ability to regulate against a cold (or hot) stress upon destruction of the appropriate nucleus, and effector signals may be evoked upon thermal or neural stimulation of each nucleus.

d. Experimental evidence shows that the hypothalamus receives temperature signals from most of the skin area as well as from receptors located at a variety of points within the body. In addition, the hypothalamus is believed to contain temperature receptors itself (von Euler, 1950) and is thus responsive to its own temperature, essentially that of the circulating blood. The essential role of the hypothalamus in temperature regulation makes it necessary to measure its temperature in experimental studies. This can now be done quite effectively (Benzinger and Taylor, 1963).

e. The hypothalamus appears to act upon this large amount of temperature information, and sends appropriate signals to the effector organs: the muscles, sweat glands, and the peripheral circulation. These efferent signals are probably not a unique function of the many temperature inputs to the hypothalamus but also take into account other demands upon the effector organs.

f. It appears impossible to identify *the* single regulated variable, but rather it is the temperature distribution that is regulated to maintain the overall energy balance (Hammel, 1968). See also Section 8.9.

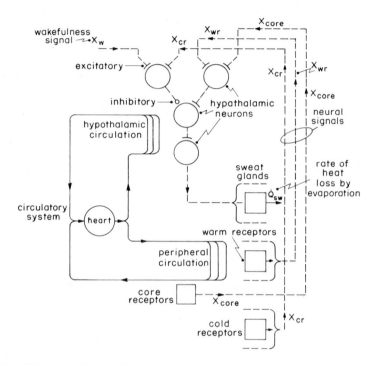

Fig. 8-5. Schematic diagram for body temperature regulation by sweating. The hypothalamic neurons and their connections are hypothetical, and simply show one manner by which they might account for the observed behavior (Hammel, 1965). The diagram corresponds with the control equation: $\dot{Q}_{sw} = \alpha_{sw} [T_h - T_{h0} - x_{cr} + x_{wr} + x_{core} - x_w]$ where T_h is the hypothalamic temperature, T_{h0} its threshold value, x_w is a wakefulness term, and α_{sw} is a proportionality constant. Suitable weighting factors for each of these input signals to the hypothalamus are not shown.

a. The regulation of body temperature under a heat stress is accomplished principally by control of the sweat rate whereby heat is removed from the body by evaporation from the skin surface. The sweat glands (effectors), as well as the receptors, are widely distributed, and it is the hypothalamus that serves to integrate and coordinate their individual actions.

b. It would be a practical impossibility to detail the individual behavior of all these receptor and effector units, together with their integration within the hypothalamus. To circumvent these myriad details, many suggestions have been made regarding the form of a control equation; a particular example being shown in this figure. This equation shows in a general manner the way in which the effector (control) action depends on the receptor signals, but inasmuch as these inputs (x_{cr}, x_{wr}, etc.) represent only some average value of the signals from a large number of similar receptors, all information regarding the temperature distribution is lost.

c. The neurons shown in the hypothalamus follow the suggestions of Hammel (1965), who has postulated that units with a positive temperature coefficient together with inhibitory neurons could account for the control action. Some experimental evidence supports this assumption (von Euler, 1950; Eisenman and Edinger, 1971).

d. The wakefulness term x_w, which is positive when the animal is awake, is introduced to account for the sudden increase in heat losses on falling asleep.

e. Identification requires that one establish both the necessity and the transmission properties of the components and pathways. Surgical intervention, either by sectioning a pathway or ablation of a component, is required to establish the necessity, and thermal or neural stimulation is used to determine the transmission. However, stimulation at any point in an intact animal produces a response due not only to the direct stimulation, but also to the signals fed around the loop. That is, as long as the feedback loop is intact, the observed response is that of the entire system and the properties of one or a group of components can not be isolated. This matter is discussed in greater detail later.

f. This regulating system includes all three of the types of biological fluxes mentioned in Section 2.17. Insofar as analysis of the regulating system is concerned, each of these fluxes is simply a signal represented by an appropriate variable. Differences in the physical character of these fluxes are of no consequence in the analysis of dynamic behavior.

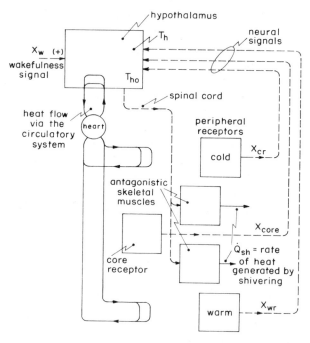

Fig. 8-6. Temperature regulation by shivering. The control equation is assumed to be of the form: $\dot{Q}_{sh} = \alpha_{sh}\,[T_{h0} - T_h + x_{cr} - x_{wr} - x_{core} + x_w]$ where $\dot{Q}_{sh}$ is the rate of heat production by shivering (above that attributable to normal metabolic processes), α_{sh} is a constant, T_h is the hypothalamic temperature, T_{h0} is its threshold value, and x_{cr}, x_{wr}, x_{core}, and x_w are neural signals as indicated (Hammel *et al.*, 1963).

a.   In contrast to sweating, which acts to increase the heat losses, shivering increases the heat generated within the body to counteract the fall in internal temperature. The block diagram shows the principal pathways and components.

b.   The hypothalamus receives two kinds of signals as in the previous diagram; the temperature of the circulating blood, and neural signals from the receptors. The hypothetical neural circuits shown in Fig. 8-5 are omitted here; presumeably similar circuits for shivering could be constructed.

c.   Although signals for the control of shivering pass down the spinal cord, the shivering frequency is established peripherally, possibly in the spinal reflexes associated with all skeletal muscles. Shivering brings about a cocontraction of antagonistic muscles, so that although there is strong contraction and heat is produced, no external work is done. There might be a temptation to ascribe shivering to an instability in the stretch reflexes, but this explanation proves to be inadequate (Stuart *et al.*, 1963). Nevertheless, the minimal damping found in the stretch reflex would be conducive to the maintenance of shivering under the proper forcing conditions.

d.   The control equation fails to reveal that $T_h < T_{h0}$ is necessary for shivering, regardless of the peripheral temperatures. Cooling of the limbs leads to shivering, but only after a minute or so, presumeably the time required for circulating blood to reach the hypothalamus. In contrast, sweating occurs within a matter of seconds after warming the extremities.

e.   Sign inversion must occur, by virtue of the fact that thermogenesis takes place as a result of a fall in temperature. A more detailed examination of the cool receptors makes it clear that the sign inversion is attributed to the negative slope of the receptor characteristic.

f.   Shivering is a graded process, starting as an increase in muscle tone, and culminating in the almost maximal cocontraction of antagonists. Furthermore, shivering frequently starts in the more peripheral muscles and gradually spreads to more central ones as the cold stress increases. One notes that the shivering muscles are ordinarily those not required for other tasks at the moment, so that there is considerable plasticity in the effector response to cold.

g.   Temperature regulation is generally accepted as being proportional, that is, some displacement of the central or peripheral temperatures is required to maintain a given control action. Inasmuch as these receptors are tonic units, they are suitable for use in a proportional regulating system.

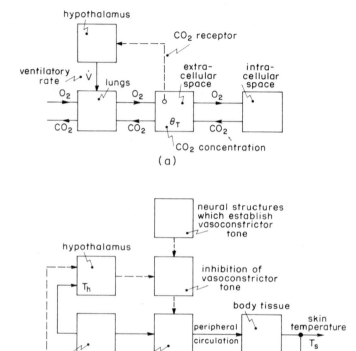

Fig. 8-7. Processes which provide an inherent sign reversal. (a) Control of respiration. The ventilatory rate ($\dot{V}$) increases with increase in concentration of CO_2 in the tissues (θ_T), whereas an increase in $\dot{V}$ decreases θ_T by removing CO_2 at a greater rate. (b) Control of peripheral circulation. Arteriole resistance is maintained by vasoconstrictor signals, which in turn may be inhibited by signals arising in skin and hypothalamic thermoreceptors.

a.   The sign reversal required in a negative feedback system may take place biologically by some process in which the output variable decreases as the input signal increases, and vice versa. This is an essential function within a negative feedback system, but is seldom performed physically in the manner suggested in Fig. 4-4a. The canonical block diagram implies a simple subtraction; reference input minus the regulated variable. Although this is a satisfactory functional statement for some purposes, the actual physical form may appear to be quite different.

b.   In body temperature control under a heat stress, the sweat rate increases and more heat is dissipated (Figs. 4-12 and 8-5). This increase in the loss rate serves to reduce the heat content of the body, and reduce its temperature. The sign reversal required for negative feedback is implicit in the control of the loss function.

c.   A somewhat similar situation exists in the control of respiration wherein the ventilatory rate $\dot{V}$ (tidal volume flow per minute) increases with an increase in tissue CO_2 concentration (part (a) of the figure). Respiration may be viewed as a means of tissue CO_2 regulation, and the sign reversal occurs because of the increased loss rate of CO_2 that follows any increase in $\dot{V}$ (Grodins *et al.*, 1954).

d.   In the two previous examples (temperature regulation and respiration) there was an increase in loss rate in order to decrease the energy (heat) or material (CO_2) stored within the body. In the case of thermogenesis associated with shivering (Fig. 8-6) the sign reversal takes place at the cold receptors themselves, the negative slope of which provides an increase in afferent neural signal when the temperature falls.

e.   Part (b) shows vasomotor control of the peripheral circulation. Forebrain structures normally maintain a maximum vasoconstrictor tone which keeps the arteriole resistances at a maximum value, and the peripheral blood flow at a minimum. Neural signals from the hypothalamus serve to inhibit the vasoconstrictor signals and permit increased blood flow under a heat stress. However, it will be seen that if vasodilatation occurs as a result of signals from the warm receptors, there are two sign reversals, one with inhibition of vasoconstrictor signals and the second from the increased circulation following a reduction in vasoconstrictor tone. The feedback in this pathway is thus positive, with the likelihood of oscillations in peripheral circulation. This would not be the case if the control were via hypothalamic temperature alone. The precise relations in this control system are apparently unknown (von Euler, 1961).

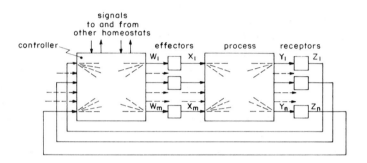

Fig. 8-8. Generalized block diagram constructed to show the multivariable aspects of a homeostat. The large number of effector and receptor organs clearly indicate a multivariable system. However, these variables are strongly coupled to each other in the controller (CNS), and the process itself, as is implied by the dotted internal connections suggested within each of the boxes.

a.   This block diagram is an attempt to generalize features that appear in many homeostats. In the case of temperature regulation by shivering one can make the following identifications. The receptors include units located in the skin, together with those in the hypothalamus, and probably at other points within the body. The controller consists of that portion of the central nervous system (CNS) concerned with temperature regulation by shivering, and centered in the posterior nucleus of the hypothalamus. The effectors are the skeletal muscles. The process is the entire body having both heat sources and sinks (the environment), and through which there is a continuing heat flux.

b.   Although each thermal receptor sends its own signal to the hypothalamus, these many signals are "integrated" by the CNS to produce the necessary control signals for the effectors. The term "integration" is here used in a general, nonmathematical sense to indicate that the neural signals leaving the hypothalamus are functions of all, or a large number, of its input signals. The hypothalamus produces a set of control signals for the effector units which reflect the overall state of the organism.

c.   In the case of shivering, the effectors (muscles) are the sites of controlled thermogenesis, and with changes in these heat sources there will be corresponding changes in the entire heat flow pattern in the body. Thus one should view the body tissues as another "integrative" process in which heat from many discrete sources sets up a heat flow pattern and temperature distribution that reflects all the thermal sources and sinks. The body serves to integrate these relatively discrete units and produce the temperature distribution sensed by the receptors.

d.   The above discussion provides little help in the identification of a regulated variable. Although the temperatures of the extremities may differ by a number of degrees from that of more central regions without lethal effects or even serious impairment of function, this cannot be said of the heart or CNS, where changes in temperature of a degree or so may have irrevsible effects. Thus the brain, and especially the hypothalamus, may be considered as the regions in which temperature regulation is the most important, although even there the designation of a single temperature as the regulated variable can be misleading.

e.   This discussion should also serve to reinforce the idea that there is probably no one reference input that alone establishes the operating point. Instead, the operating point is set by the joint action of a large number of constant terms appearing in connection with many of the individual processes.

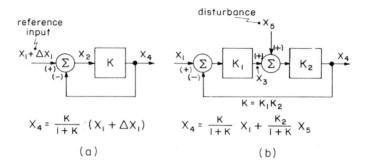

reference
input

$X_1 + \Delta X_1$ $\xrightarrow{}$ Σ $\xrightarrow{X_2}$ $\boxed{K}$ $\xrightarrow{X_4}$

disturbance $\rightarrow X_5$

X_1 $\xrightarrow{}$ Σ $\xrightarrow{}$ $\boxed{K_1}$ $\xrightarrow{(+)}$ Σ $\xrightarrow{}$ $\boxed{K_2}$ $\xrightarrow{X_4}$

X_3

$K = K_1 K_2$

$$X_4 = \frac{K}{1+K}\,(X_1 + \Delta X_1)$$

$$X_4 = \frac{K}{1+K}\,X_1 + \frac{K_2}{1+K}\,X_5$$

(a)   (b)

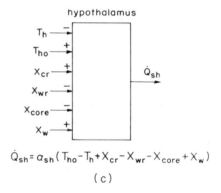

hypothalamus

T_h $\longrightarrow$ —
T_{ho} $\longrightarrow$ +
X_{cr} $\longrightarrow$ +
X_{wr} $\longrightarrow$ —
X_{core} $\longrightarrow$ —
X_w $\longrightarrow$ +

$\dot{Q}_{sh}$

$$\dot{Q}_{sh} = a_{sh}\,(T_{ho} - T_h + X_{cr} - X_{wr} - X_{core} + X_w)$$

(c)

Fig. 8-9. Changes in the steady-state operating point may be produced by a change in the reference input as in (a), or by an additive disturbance as in (b). (c) The block diagram and postulated control equation suggests the quantities upon which the rate of shivering depends. Note that suitable weighting factors for each of the inputs are not shown; the inputs are not of equal significance in determining $\dot{Q}_{sh}$.

a. The discussion in Chapter 4 and reference to Figs. 4-11 and 5-4 show that in a strictly linear system, a change in the set point or reference input produces a proportional change in the magnitude of the regulated variable at the steady-state operating point. If the loop gain is sufficiently great, the two changes are nearly equal. This is shown in (a) of this figure.

b. The additive disturbance shown in (b) also changes the operating point, and the steady-state change in x_4 is proportional to the disturbance x_5. However, without considerable knowledge regarding the system and its components, it is impossible to distinguish between these two cases simply by observing the consequences for x_4.

c. Our interest in the above relations stems from the desire to properly account for changes in the regulated variable ("body temperature") under a number of conditions. In Fig. 8-1, a sustained temperature change was found for each work load, and the question arises as to whether this is a change in the reference input or simply a loading effect. The effectiveness of the regulating system is undoubted, and our only interest is in a satisfactory representation. A similar question arises in fever and hibernation, where sustained temperature changes are also observed (von Euler, 1961).

d. Consider the case of shivering and the postulated control equation given in (c). The hypothalamus is assumed to have six input signals that are appropriately weighted (not shown) in determining the efferent signal $\dot{Q}_{sh}$. Note, however, that of the input signals four are dependent variables (x_{cr}, x_{wr}, x_{core}, and T_h) and therefore subject to change with the loading on the system. The other two inputs (T_{ho} and x_w are probably independent of loading, and thus may be properly considered as constituting the reference input. Although a change in any of the quantities in the control equation has been said to produce a change in the set point (Hammel, 1968), a refinement in this terminology may be well advised in order to distinguish between a change in the reference input, and a change due to loading effects in the system.

e. The threshold value for hypothalamic function T_{ho} and the wakefulness signal x_w might serve as reference inputs if it can be shown that they are truly constants that are independent of the body temperatures. The other inputs to the hypothalamus, all functions of the body temperature, are simply feedback signals and can hardly be considered as contributing to the reference input.

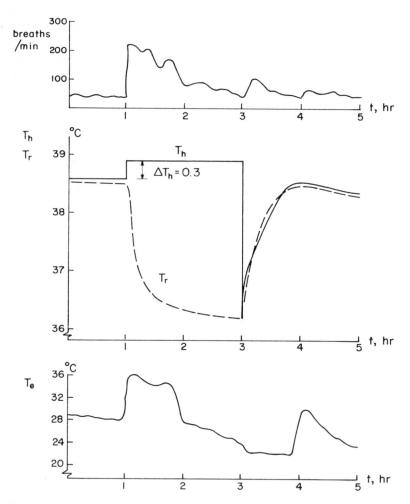

Fig. 8-10. Disturbance to the temperature regulating system of a rabbit. The hypothalamic temperature T_h was "clamped" at 38.9 °C by a controlled thermode. The rectal and ear temperatures (T_r, T_e) were measured and the above curves sketched from the experimental data (von Euler, 1964).

a. The previous chapters have provided ample evidence for the importance of loop gain in the behavior of a feedback system, and inferentially have suggested the necessity of knowing the gain magnitude in order to develop an adequate understanding of a homeostat. Section 4.5 showed that for the linear, single-loop unity feedback regulator, it would be possible to calculate the loop gain from a measurement made on the intact system (a closed-loop measurement). For any other case additional measurements are required.

b. In a physiological system, the direct measurement of loop gain, that is, with the loop open, presents many problems. In the first place, the integrity of the components must be maintained, and the operating point at which the gain is measured must be the same as with normal closed-loop operation. Second, a basic difficulty in opening the loop in a physiological system arises from the existence of multiple pathways and the necessity of opening all paths for a true open-loop measurement.

c. An experiment in gain measurement in the body thermostat is summarized in this figure. The hypothalamic temperature was elevated slightly by a stimulating thermode, which served to "clamp" the temperature in a small region. As a consequence, the peripheral circulation increased, as indicated by ear temperature, and the rectal temperature T_r fell. The ratio $\Delta T_r / \Delta T_h$ in the steady state was interpreted as the loop gain; it had values of 7.6 to 9.7.

d. Examination of the block diagram in Fig. 8-7 will show that although clamping T_h had the effect of opening the feedback path provided by the circulatory system to the hypothalamus, the neural feedback pathway from the warm receptors was still effective. The total input to the hypothalamus was therefore the thermal signal ΔT_h plus an additional neural input from the warm receptors, although the precise contribution of the latter cannot be assessed in this experiment. The true loop gain cannot be calculated in the manner suggested above, but would probably be somewhat less than the values given.

e. In addition to this difficulty, there are other aspects of the experiment that throw doubt on the validity of this measurement of loop gain. Although the hypothalamic temperature was clamped with a thermode, the temperature of which was carefully maintained, there is some question as to whether this temperature was actually that of the local receptors. In addition, the gain measured was that from hypothalamus to rectal temperature, and did not include the remaining portion, namely, the transmission from rectum to the hypothalamus.

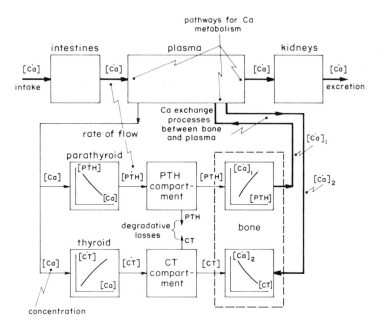

Fig. 8-11. Regulation of Ca concentration in the blood plasma. Principal pathways for Ca metabolism are shown from food intake to urinary excretion. Exchange of Ca with bone is shown as the sole control mechanism. However, control of Ca absorption in the intestines and resorption in the kidneys are also possible, but believed to be of secondary importance and are not shown in the figure. Hormones PTH and CT have opposite effects on Ca exchange with bone; whereas PTH promotes the movement of Ca from bone to plasma, CT serves to increase movement of Ca from plasma to bone.

Regulation of the Ca Concentration in 8.12
Blood Plasma: Fig. 8-11

a. Calcium regulation is selected as an example of homeostasis as applied to the concentration of a specific chemical species. Regulation is necessary because both the intake and the demand for Ca may be subject to wide variations, while at the same time small variations in Ca level can have serious effects. In this example, a store of Ca is maintained in the bones and can be made available to the plasma when required, and replenished when it is possible to do so.

b. The normal level of Ca in the plasma of man is 10 ± 2 mg/100 ml, and its presence is essential for life. Hypocalcemia leads to greatly increased excitability of nervous tissue and muscle cells (among other things), which in the limit produces extreme tetany of all muscles and death by respiratory paralysis. Hypercalcemia, although not as likely to occur, can result in kidney stones and other disturbances.

c. The calcium level is regulated by means of hormones secreted by the parathyroid, and the C-cells in the thyroid. These two glands are closely associated anatomically, and this fact led to confusion in the early extirpation studies until it was realized that the precise surgical method employed played a dominant role in the experimental findings (Copp, 1969). It is now agreed that the parathyroid gland secretes *parathormone* (PTH) at a rate that varies inversely with the plasma Ca level. Specific cells within the thyroid gland secrete another hormone *calcitonin* (CT), but its rate of secretion varies directly with the Ca level. The terms calcitonin and thyrocalcitonin both appear in the literature, but they refer to the same substance, with calcitonin now accepted as standard.

d. PTH has three effects within the body: (1) it increases the rate of exchange of Ca from bone to plasma, (2) it decreases the rate of Ca excretion from the kidneys, and (3) increases the rate of absorption of Ca in the gut. Thus PTH acts at three or more sites (the last two are not shown in the figure) and in all instances in a manner to increase the Ca level in plasma.

e. CT exerts its principal action on the exchange process between bone and plasma where it increases movement of Ca from plasma to bone. The evidence that CT may also affect the excretion rate is controversial.

f. Both hormones are present in plasma at normal Ca levels, but their effects are antagonistic in that PTH serves to protect against low Ca, whereas CT protects against hypercalcemia.

g. The response of the thyroid–CT system is fast compared to that of the parathyroid–PTH system. Further examination of the dynamics of this regulatory process may help to explain the contributions of each of these hormones in more detail.

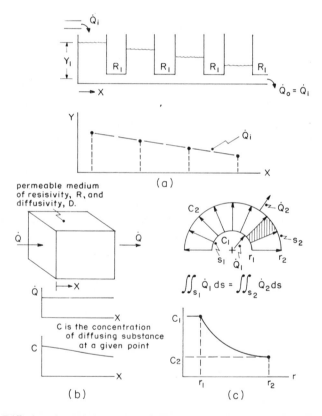

Fig. 8-12. Diffusion through lumped and distributed parameter processes in the steady state. (a) Flow through four identical compartments. (b) Linear flow through a uniform medium. $\dot{Q} = -(1/R)(dC/dX) = -D(dC/dX)$. (c) Radial flow through the wall of a cylinder showing the distribution of concentration along a radius.

a. Most of the examples in previous chapters have been compartmental models similar to the one shown here in (a), in which material or energy flowed from one compartment to another at a rate proportional to the difference in concentrations, and inversely proportional to some resistive property of the medium. Such a *lumped parameter* or *compartmental model* is adequate if one can identify regions of uniform concentration.

b. A more general situation occurs when flow takes place through a region in which the concentration varies from point to point, and the "resistive" property of the medium is distributed rather than being concentrated in a few locations as in (a). The simplest case, as suggested in (b), is one in which the flow lines are parallel; it is termed *one-dimensional diffusion,* and the model is said to have *distributed parameters.*

c. Many flow processes are described by *Fick's law of diffusion* (Riggs, 1963), which states that the flow at a point, $\dot{Q}$, is proportional to the *concentration gradient, dC/dX*. Note that gradient is the counterpart of concentration difference, expressed on an infinitesimal basis, so that Fick's law is essentially the same as the relations previously applied to compartmental flow except that it relates to the conditions at a point. The constant of proportionality between flow and concentration gradient may be expressed as a resistivity R, or a diffusivity D, where $D = 1/R$.

d. Diffusion through a given region may be visualized as the flow through a large number of compartments proportioned so as to preserve the lines of flow. Specifically, in (b) the permeable medium can be replaced by a very large number of compartments and flow resistances in the manner shown in (a). In this case the compartments and resistances would be of equal magnitude with the total volume and resistance held constant. The multicompartment model becomes a better approximation of the process as the number of compartments increases.

e. In part (c) of the figure, the diffusion of particles (or heat) takes place in a radial direction through the wall of a hollow cylinder. If the cylinder has uniform properties, then in a lumped parameter model, the compartments would increase in size and the resistances would decrease toward the outside. The latter occurs because the cross section to flow increases toward the exterior. For given concentrations at the two bounding surfaces, the distribution of concentration within the cylinder adjusts itself somewhat, in the manner shown. Although the flow per unit area decreases toward the outside, the total outward flow through the cylinder in the steady state is the same at all radial distances.

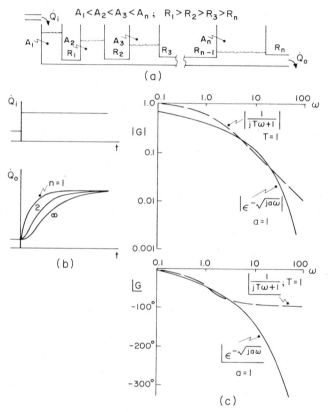

Fig. 8-13. (a) Lumped parameter model for diffusion through the sector of a cylinder, shown shaded in Fig. 8-12c. (b) Responses to step function inputs for systems of first, second, and infinite order (diffusion). (c) Spectrum of a specific one-dimensional diffusion process whose sinusoidal transfer function is $\exp[-(ja\omega)^{1/2}]$, along with that of a single compartment $[1/(jT\omega + 1)]$. For comparison purposes, $a = T = 1$.

a. It was suggested in the previous section that the diffusion of particles or heat through a given medium may be approximated by representing the process as a compartmental model; the accuracy of the model improving with an increase in the number of compartments. Diffusion through the wall of a cylinder, as shown in Fig. 8-12c, may be approximated by the compartments shown in (a) of this figure. Inasmuch as the diffusion is along radial lines, this series of compartments represents the flow through a sector of the cylinder, and the approximation would be improved if the size of the individual compartments were made smaller as their number was increased.

b. In the earlier discussions of dynamic behavior of compartmental models, it was shown that the order of the system was, in general, equal to the number of compartments, and that the number of transient terms in the solution of a dynamic problem was also equal to the system order. These concepts carry over to the process of diffusion, and lead to the conclusion that it is of infinite order, so that the solution to a dynamic problem will contain an infinite number of terms. It is reasonably obvious that the methods previously applied to compartmental models are not adequate here, except as they provide acceptable approximations to a complete solution.

c. The mathematical expression for the solution to a diffusion problem is frequently, but not always, a series of decaying exponential terms of the same form as those encountered earlier. The complete solution may be approximated in two different ways. Many of the terms in a complete solution become vanishingly small in short intervals of time, so that although there are an infinite number of components, they are not of equal weight and many may be discarded. On the other hand, the diffusion process may be approximated by a lumped parameter model, as suggested above, and the solution of the latter used to approximate the true solution. These two approximations are not the same.

d. Some appreciation of the consequences of replacing the original diffusion process by a compartmental model may be gained from the figure in (b), in which the step function responses of three systems are contrasted. These curves represent the transmission through systems in which the same total capacity (volume) and resistance have been lumped in one, two, and an infinite number of compartments. The second two curves are only approximate; their precise form depends upon the manner in which the compartments are proportioned. For all systems of order

8.14   THE DIFFUSION PROCESS   275 two and higher, there is an inflection point, but one does not observe any qualitative change in the character of the S-shaped curve as the order increases beyond two.

e. Despite the fact that diffusion is described as a process of infinite order, if it obeys the laws previously given, it is still linear and responds in a sinusoidal manner to stimuli of the same form. It is thus possible to describe the process with a frequency spectrum as in (c), where there is also shown the spectrum of a first-order process (a single compartment) for comparison. In the latter case, the volume is concentrated in one compartment with one resistive outlet, whereas in the distributed parameter case the same volume and resistance are subdivided among an infinite number of compartments.

f. A distinctive aspect of the diffusion process is the large value of phase that appears at high frequencies. As a matter of fact, for any diffusion process the phase angle will increase without limit as the disturbing frequency is made larger, and the phase does not approach some asymptotic value as in compartmental models. Although the transient response as shown in (b) gives no evidence of any qualitative difference between the lumped and the distributed parameter cases, the spectrum shows very clearly that the maximum phase angle with diffusion can become infinitely large. This is never possible with a lumped parameter system of finite order.

g. An examination of the stability of systems containing diffusion shows that the potentially large phase angles may lead to instability, as was the case with the time lag process discussed in Chapter 7. The ubiquity of diffusion as well as transport lag throughout the body strongly suggests that the contribution of these processes to the stability (or instability) of all homeostats should be examined carefully. Furthermore, possibly the greatest danger in using any lumped parameter approximation is in its failure to properly predict the stability limit of the prototype.

h. It is tempting to compare the dynamic properties of diffusion and time lag, in particular as they effect the stability of a feedback system. In both instances the phase angles can become greater than those attainable with lumped parameter models. Time lag introduces no attenuation, whereas diffusion does attenuate the high-frequency signals. In a sense, then, time lag might be said to have a greater destabilizing effect on a given system than diffusion, but it is doubtful if this comparison has anything other than a very limited significance in duscussing a given system. Of greater importance is the fact that a lumped parameter model of either of these two distributed parameter processes may lead to completely erroneous conclusions as regards stability. This observation assumes considerable importance when it is realized that a full picture of the factors contributing to the stability of most homeostats is still not available.

i. This discussion of diffusion has without doubt obscured the fact that precise calculations of any problem introduces mathematical complexities that only a computer can handle. Although the relations described above may be valid in a qualitative sense for many diffusion processes, when it comes to developing the correct quantitative relations for a specific problem, one finds that there are no general solutions but rather a host of specific solutions each unique to the geometry of that process. This is because the flow pattern in diffusion is strongly dependent upon the geometry of the flow, and the so-called boundary conditions existing over the bounding surfaces at which the diffusion originates and terminates. The calculation in detail of a diffusion problem, for example, its spectrum or its time response to some disturbance, will in general yield an answer valid only for that geometry and set of boundary conditions. The results of analysis of diffusion are therefore a host of individual solutions.

j. The reader is again reminded that although the above discussion has been largely couched in terms of the flow of material, particles, or chemical species, most of the relations are equally applicable to the flow of heat. The application of both lumped and distributed parameter models to the study of body temperature regulation has been considered by a number of individuals (Wissler, 1964; Stolwijk and Hardy, 1966). For further references, see Jones (1969).

Homeostats and Power Supply Dynamics 8.15

In all discussions of feedback regulation up to this point, there has been no mention of energy or power, and yet a cursory examination of any of the systems described will show that power is not only necessary for their operation, but also intimately related to their behavior. This is true for the fluid flow systems described earlier, and in fact for any technological regulator. In all cases, not only does the regulator in a sense control the flow of energy, but it must use energy in the control process.

A similar observation may be made for homeostatic systems in that regardless of the physical quantities being regulated (movement, temperature, chemical process), energy in an appropriate form is needed. In living systems, the immediate form of energy is found in the ATP bonds, and this chemical energy must not only be available but transformable as needed to provide power for the processes and reactions required for regulation (Green and Goldberger, 1967).

Despite this need for energy to carry out the various regulatory actions, neither energy nor power have appeared explicitly in the equations that have been formulated to describe these systems. The equations have been in terms of a variety of physical variables (velocity, fluid flow, position, temperature), but at no point has energy been explicitly used. It will be found quite generally that the operating characteristics of a dynamic system can be described completely using physical quantities other than energy. This means that such properties as sensitivity, transient response, stability, and the frequency spectrum do not depend upon energy directly, but rather upon a set of physical laws that in turn are derived from the conservation of mass and energy. This is not to say that the equations for a regulator, for instance, could not be written using energy or power as a variable, because this is quite possible. However, it is not done in most cases simply because to do so would make the equations more complicated.

This leaves us with the question: does not energy or power level affect in some manner the behavior and properties of a dynamic system? The direct answer is that it does, and the effects are implicit in the equations we have been discussing. With an increase in the energy level, the size of the components increases and more energy and material is stored within the system. The energy level is thus reflected directly in the size of the compartments, as well as in the magnitudes of the several variables. For example, the roots of the characteristic equation vary inversely with the time constants of the system and will therefore change with the physical size of the components.

The above discussion points to the fact that energy appears only implicitly in the equations of motion, and then usually as related to certain of the parameters and constant terms appearing therein. While this is the usual case in a technological system, there is an interesting exception that may be of physiological import. If the source of energy is unable to provide all the power required at each instant of time, then the dynamics of the power supply become a factor in the overall system behavior. From a slightly different viewpoint, the dynamic analysis of any system must consider all components that are coupled together, the coupling being such that a disturbance at any point in the system will cause all dependent variables to react.

An interesting example in which power supply dynamics are crucial to an understanding of system behavior is that in which two regulating systems obtain their power from a common supply. If that supply is somewhat short of being adequate, there is in effect a competition between the two regulators for the power available. In the total system behavior, not only will the transients due to the disturbances on each individual regulator appear, but the disturbances on one will evoke responses on the other. The two regulating systems, ostensibly independent units, are in fact but parts of a greater whole, being coupled together by the common power supply.

There is little question but that the many homeostats in a single organism are interconnected in a variety of ways, although a thorough study of these possibilities remains to be carried out. One mode of coupling could certainly be through the common energy supplies, those molecules whose breakdown serves to release the energy required for operation of the homeostats.

Adaptation and Adjustments 8.16
in Homeostats

In addition to the immediate response of the control components in a regulating system, there are in many homeostats a number of slower-acting processes, the effect of which is to improve the quality of the regulation. Although the terminology regarding such "adaptive" changes has become confused and confusing, we shall attempt to avoid some of the resulting difficulty by defining terms in a somewhat unconventional manner. The regulating system or homeostat reacts to a variety of disturbances by bringing into play one or more control actions, and doing so in a reasonably rapid manner. The rapidity is limited by the dynamics of the processes themselves. In contrast to these control actions, there is a category of other responses that may appear when the disturbance or stress is relatively long lasting; these serve to reinforce, modify, and sometimes replace the immediate control action. We shall denote such a slow change as an *adjustment*. It will be seen immediately that the principal distinction between the two response categories is that of response time. The regulatory behavior takes place "immediately" and the adjustments require longer periods to develop, and also to disappear. The use of response times to separate these two categories is obviously imprecise, but appears to be a reasonable procedure.

The term adaptation, though widely used in this context, has unfortunately acquired two quite different meanings. In the biological literature *adaptation* is used to describe the changes in structure, function, and behavior that are genetically determined, having been acquired over long periods by evolutionary processes. This usage has strong sanction from biology and recognizes that each species has become adapted to its own environment in a unique manner. *Adaptation* is also used to describe those changes that occur within a single lifetime, are reversible, and presumably are of a character to enhance the fitness of an individual for his own environment. This latter usage is somewhat clarified by terming it *physiological adaptation*. However, we propose to employ the word *adjustment* for this category, to designate those changes that take place in a homeostatic system when under a protracted stress, it being understood that an adjustment will also disappear with time when the stress is removed.

The literature is also confusing when homeostatic systems are described as being adaptive. It would be more accurate to say that homeostats are adapted, having acquired their adaptation by evolution. This same homeostat can properly be said to be adjustable only if it has the ability to further modify its behavior with continuing stress.

Adjustments are sometimes divided into two categories. *Acclimation* denotes those adjustments that occur as a consequence of a change in one physical quantity in the environment, such as temperature acclimation. A somewhat broader term *acclimatization* is reserved for adjustments induced by any combination of environmental factors. It describes a broad range of adjustments, a total reaction of the organism to environmental change (Prosser, 1964).

The terminology suggested above attempts to reserve the term adaptation and its derivatives for evolutionary processes. Despite this attempt at clarification, the development of somewhat similar concepts in engineering technology has further strained the situation. A class of technological regulators has emerged, currently described as *adaptive control systems* (ACS), which have, however, a distinctly different mode of operation from that observed in physiological counterparts. The ACS is usually found to operate in conjunction with some optimal principle relating to a property of the system such as speed of response, energy consumption, or error. The quantity to be optimized, and the specific optimization format, are selected by the design engineer, and suitably incorporated into the system design. In one form this may require a computer programmed to continually calculate some performance index, and then execute the necessary control actions to realize optimum performance. The engineer, as the designer of the system, selects those optimizing principles that appear to meet the performance specifications. Evolutionary aspects do appear in engineering, but only as each successive design is presumeably an improvement on the previous generation. The ACS would be termed an *adjustable control system* if one adhered to the definitions proposed above.

Turning to biological systems, it is clear that the situation is quite different. In the first place, optimal performance can only be interpreted in an evolutionary sense, and an optimal principle should be directly related to survival of the species. From this standpoint, it makes little sense to consider optimization from a physiological standpoint alone, although clues may be obtained from observation of a single organism. Although one may accept the premise that evolution results in the optimization of survival, the translation of this notion into physiological terms is far from clear. It is noted, however, that adjustments may serve to broaden the range of stress conditions under which regulation is maintained, and that any modification of a homeostat to promote economy would probably assist survival.

In addition to the previously described control actions (shivering and sweating) it is not difficult to find adjustments that reinforce these actions when the thermal stress is prolonged and intense. In the case of a cold stress, while shivering can restore and maintain thermal balance under many conditions, additional thermogenesis appears as the result of a general increase in metabolic rate. The endocrine system, most probably the adrenals and the thyroid gland, bring about an increase in body metabolism apart from any shivering activity. This hormonal control of heat production, which develops slowly, has been demonstrated in animals and presumeably also occurs in man although experimental evidence on this point is not complete.

A second example of an adjustment is found in the sweat glands, where following prolonged activity the sweat rate increases for the same neural innervation. Although there is recruitment of sweat glands as the heat stress increases, the adjustment appears to consist of an increase in "gain" of the glands; they become more effective producers of sweat for evaporation. Again, this effect develops over extended periods of heat exposure (days) and is lost over a similar period upon return to normal thermal conditions.

It may be inferred from this discussion that adjustments to a regulatory system in effect lead to a new system for which a new model and new analysis may be necessary. Furthermore, there immediately arises many questions of experimental protocol in that observations on the system are of little value unless the state of adjustment (adaptation) is defined and maintained throughout any series of experiments. Adjustments have been defined as those regulatory changes that occur as a consequence of prolonged disturbances, so that experimental studies should be carried out with disturbances of relatively short duration and with care that the average values of the many variables are maintained relatively constant.

This requirement alone is one of the reasons why sinusoidal testing is proving to be so attractive. A stimulus of this character serves to maintain the adjustive state of the organism and yet produces a disturbance (a continuing sine wave) that is entirely adequate to reveal the dynamic properties.

Disturbances to Biological Systems   8.17

The previous section attempted to emphasize that adjustments to homeostats may take the form of changes in practically all aspects of the regulating system. A similar observation may be made with regard to disturbances imposed upon the system; they may affect any or all of the attributes of the system. The essential point is that a disturbance (or an adjustment) will usually result in a new system, one whose describing equations are in some manner different from the initial set. The disturbance may be as deceptively simple as a change in the numerical value of a coefficient or as extensive as a wholly new equation, but the net effect on system performance can be drastic in either case. The only tenable conclusion is that any disturbance may lead to an entirely new "ball game."

It is sometimes helpful to group possible disturbances into three categories: (1) input disturbances, (2) parametric changes, and (3) output loading. An *input disturbance* refers to any factor that affects the inflow of energy, material, or information into the physiological system. In the fluid flow systems of earlier chapters, the flow into any of the compartments from outside the system may be subject to disturbing factors in the environment. In tracking a moving target with the eyes, the input is the image of that target on the retina, an obviously changing input quantity. The input of food and water, being highly variable quantities, constitute input disturbances for a number of homeostats.

Parametric changes are represented in fluid flow systems by changes in compartment size or in the resistance to flow. The latter might occur as a consequence of a variation in the resistance parameter itself (the constriction) or a change in viscosity brought about by chemical or physical changes in the fluid. Respiratory diseases (fibrosis, asthma, emphysema) may affect both the air flow resistance and the lung capacity. Since the ultimate function of respiration is to deliver O_2 to the tissues and remove CO_2, any changes in blood circulation may well upset these two processes. This is an example in which the coupling between two homeostats may be viewed as a disturbing factor on one of the regulating systems.

Load or output disturbances occur when there is a change in the demand for material or energy leaving a regulated system. The mechanical load on a muscle is given by its length, velocity, and force so that changes in posture or body activity may be considered a load disturbance. Changes in the temperature or humidity of the environment, or in the clothing worn by the subject, constitute load disturbances for the body thermostat.

It is readily seen that the above classification is a rough one and many borderline cases will occur to the reader. Another set of categories might be constructed on the basis of the effects produced in the set of equations describing the mathematical model. The principal justification for any of these categories lies in the fact that they underscore the variety of disturbances besetting all homeostats. This concept cannot be overemphasized in view of the fact that most textbooks limit the discussion of disturbing factors to a few obvious cases, and do not reveal that every aspect of the homeostat is itself subject to change.

The foregoing discussion of disturbances may to a large extent be paralleled by a similar treatment of adjustments in that both are ubiquitous, and their consequences may be pervasive. However, a distinction between these two terms does appear when one considers the consequences of the changes. A disturbance is generally assumed to have detrimental effects as regards the quality of the regulation, whereas an adjustment is beneficial, in that it acts to preserve the quality of the regulation. In engineering technology, the term *compensation* describes changes of the same character as those here termed adjustments; they are consequences of environmental changes that tend to offset the immediate effects of a disturbance to the system. The increase in blood viscosity in the cold is believed to be a significant factor in restricting peripheral circulation under such conditions, and therefore it may be regarded as an adjustment that aids temperature regulation of the body.

The term *autoregulation* (or *self-regulation*) appears quite frequently in the literature of homeostasis as well as in engineering, but unfortunately means quite different things in the two literatures. In engineering, consider the class of systems which in the absence of feedback exhibit no tendency to reach a specific steady state; they are astatic. The principal example is that of a motor element, with the position of its shaft as the variable of interest. In the absence of an input voltage, the motor will be stationary, and its shaft may occupy any of its possible angular positions. In the terminology of Section 4.15, this is a Type 1 system, and when provided with feedback will then take up a shaft position corresponding to the input signal. This system in the open-loop condition lacks autoregulation.

In the terminology of physiology, the usage is quite different (Guyton,

1971). For example, the autoregulation of blood flow in tissues describes an ability of the tissues to help maintain a normal blood flow despite disturbances that would otherwise change the flow rate. The flow rate is quite independent of arterial pressure, the arterioles and precapillary sphincters contracting and dilating to bring this about. It is believed that this is a purely local process in which metabolites in the tissues act directly upon the muscular coat of the vessels; it is not controlled by signals from the CNS. Similar behavior is observed in many organs that are themselves able to regulate their own blood supply locally.

Examination will show that some of the processes termed autoregulation satisfy the definition of a feedback system, and would therefore be considered as minor feedback loops in a larger feedback regulating system. On the other hand, some of these processes do not satisfy the definition of feedback, do not display the properties expected of a feedback system, and are more nearly compensatory in character.

Further Discussion of Physiological 8.18
Signals and System Complexity

One aspect of system complexity was indicated in Fig. 8-8, which shows the multiplicity of parallel signal pathways and the fact that the individual signals in the several paths were coupled by both the CNS and the process being controlled. The multiplicity of paths is a direct consequence of the large number of receptor and effector units. This would certainly be true of any homeostat that included neuromuscular control, and probably also in the endocrine system in which the target organs are specific cells widely distributed throughout the body.

One possible conclusion from the observation that these many similar units are operating in parallel is that the redundancy so provided promotes reliability in the operation of the associated system, and thus of the organism itself. While there is undoubtedly much truth in this conclusion, it does not describe the whole picture. What may appear as redundancy of similar units also permits the recruitment of units as the signal increases, and of course recruitment is observed in practically all cases. The fact of recruitment permits a much broader range of signal to be effective than would be the case without it. There seems to be good evidence that the recruitment of a single unit (a motor unit, for example) is measurable and constitutes an effective signal increment. With anywhere from several hundred to several thousand units or parallel paths, one can show that a signal range of a thousand or more is quite possible, a value not attainable in many technological regulators because of noise limitations. This fact may well account for the excellent performance observed in many homeostats.

The term *multivariable control system* as applied in an engineering context connotes something quite different from that described above. In engineering usage, the term refers to a single process or machine in which a number of variables are to be regulated, such as the speed and torque of a motor. However, these two quantities are not independent, but are related to each other by properties inherent in the motor or process itself. Thus any control action that serves to modify one of them will affect both, and such coupling may not be desired. The engineering problem in systems of this character is frequently one of decoupling the variables, speed and torque in this instance, so that each may be adjusted separately, and each given a steady-state value independent of the other. Procedures are available for the design of decoupling circuits so as to permit independent adjustment of the process variables. It is not known whether this concept has any biological implications, but the term multivariable system implies quite different problems in biological and technological regulators.

The biological problem in connection with multivariable systems is one of determining the nature of the coupling between individual variables, since this is bound to have a strong effect on system performance. Closely associated with this question is the concept of *hierarchical control system,* briefly mentioned in Section 8.7 in connection with plasticity in the control of shivering muscles. Another example is that of neuromuscular control of posture and movement in which one can identify spinal, cerebral, and cortical levels of control, all operative on a single muscle. Hierarchical systems, sometimes termed multilevel, multigoal (MLMG) systems in the engineering literature, have been studied theoretically (Fleming *et al.,* 1964; Mesarovic, 1968). The physiological problem is one of isolating these several hierarchies and developing suitable experiments to study one level of control without unknown interference from another. While the neuromuscular control example is fairly well understood, this is not the case in other homeostats where interconnections between regulators are as yet not clearly defined.

Another aspect of physiological signals that does not appear in any of the diagrams is the fact that almost without exception the signals are always nonnegative. That is, neural pulse frequency, chemical concentration, and sweat secretion are inherently positive quantities, and negative values have no physical significance. This was discussed in a previous chapter in connection with fluid flow between compartments. Expressed somewhat differently, these and similar variables can reach the value zero, but input factors that might tend to force them into the negative region have no effect; there is complete cutoff and the variable remains at zero. If this occurs at any time, the transmission is effectively interrupted, and during the interval that a variable is forced to stay at a zero value the feedback is in effect nonexistent.

Closely related to the fact that the variables are always nonnegative is the use of two pathways to provide for signals of opposite sign. This is perhaps best seen in neuromuscular control, where the angle at each joint is controlled by a pair of muscles: the agonist and antagonist. A similar situation is observed in neural circuits, where both excitatory and inhibitory signals converge on a single neuron whose output, however, is always positive. Two signals, each inherently positive, but of opposite significance insofar as the final common path is concerned, may be summed algebraically by a common neuron. This could be nature's way of surmounting the limitations imposed by variables, all of which are positive.

The discussion in this chapter, as well as that in Chapter 4, has served to suggest many unresolved questions regarding the presence and role of a reference input in physiological regulators. The location of the operating point, that is, the values of the dependent variables in the steady state, is determined by the inputs to the system as well as the nonlinear characteristics of all components. Furthermore, the reference input has a diminished role in establishing the operating point because of the nonlinear components although this effect is less if the loop gain is high. Among the unresolved questions are the following. What physical quantities appearing in a biological system could possibly serve as a reference input in the classical sense, having the necessary long-time constancy of value, but also being adjustable to account for the observed changes in operating point in fever and hibernation, for example? Second, recognizing that biological components generally operate in a nonlinear region and that there are components in the feedback pathway, what role could a reference input play when its effect is diminished by these two factors? These and similar questions serve to indicate that a number of fairly basic problems regarding the steady-state operation of a homeostat remain to be answered, to say nothing regarding its dynamic behavior.

There is good reason for believing that threshold values, as seen in the case of the cool and warm receptors, might serve the function of a reference input. Although there is no information regarding the long-time constancy of these thresholds, an effective reference input might be formed from some average of the thresholds in a population of similar units.

This could conceivably provide the necessary long-time constancy, but does not explain how the reference quantity could be adjusted to change the operating point.

Epitome 8.19

1. Although homeostats are fundamentally feedback regulators, they are distinguishable from many technological counterparts by their hierarchial complexity, by the large number of pathways and components operating in parallel, and by the presence of both discrete and continuous signals. The study of physiological regulators is first of all an identification problem.

2. The identification of a negative feedback regulator within an organism is aided by the following circumstantial evidence: (a) the maintenance of physical quantities at relatively constant magnitudes in the face of internal loads and environmental disturbances, and (b) the location of a process that exhibits sign reversal.

3. Internal temperatures in man are regulated principally by shivering, vasoconstriction, and sweating, each of which is operative over a limited range of environmental temperatures. Although these three mechanisms are relatively independent of each other, there is coupling between the body temperature regulating systems and the other homeostats.

4. Temperature receptors are widely distributed over the body surface and at various internal points, and have the properties one would expect of a sensor for temperature regulation. However, the identification of one of these receptors as a component in a temperature regulating system poses an almost insuperable problem.

5. The temperature distribution throughout the body is far from uniform, and does not remain constant with changing thermal loads. Although it appears to be impossible to identify a single temperature as the regulated variable, there is general agreement that the hypothalamic temperature comes closest to this quantity.

6. Body temperature regulation by sweating makes use of neural signals from thermal receptors, as well as thermal signals from the circulatory system. The hypothalamus receives these many input signals and develops a set of output signals to the sweat glands that reflects the temperature picture of the entire body.

7. Shivering occurs only after the hypothalamic temperature has fallen, and with increase in the cold stress more skeletal muscles partici-

pate. The sign inversion is apparently produced by the negative slope of the cold receptor characteristic.

8. Sign reversal in physiological systems can be attained by a variety of physical processes. In respiration, it occurs as a result of the increased ventilatory rate that follows a rise in tissue CO_2 concentration, and in vasomotor control the reversal occurs by inhibition of vasomotor tone.

9. The designation of a homeostatic system as multivariable serves to denote the large number of receptors and effector units, operating more or less in parallel, whose signals must be integrated by the control mechanisms and the body processes. The existence of multiple receptors and effectors gives the system the appearance of being distributed, whereas the integrative processes prevent these parallel signals from being completely independent.

10. Steady-state changes in the operating point may be caused by changes in the reference input (if such exists), or by constant imposed loads. The distinction between these two cases is obvious in the block diagram but difficult to make from experimental observations.

11. Open-loop gain is difficult to measure in homeostats because (a) the integrity of the system must be otherwise unaffected, (b) the same operating point must be maintained, and (c) all parallel paths must be interrupted in opening the loop.

12. Regulation of Ca concentration in the plasma is accomplished by two hormones having opposite effects on the exchange of Ca between bone and plasma. Thus, the regulation makes use of an internal store of Ca to smooth out fluctuations in the input and output rates.

13. Diffusion may be looked upon as the limiting form of flow through a sequence of compartments as their number is increased without limit, the size of each compartment decreasing as the number increases. A given physical problem involving diffusion may be approximated by a compartmental model if the material in flow is but a small portion of the total material, and the flow may be assumed to take place between regions of uniform concentration.

14. The step function response of a diffusion process contains an infinite number of exponential components, and may be regarded as the response of a compartmental model having an infinite number of compartments. In the frequency domain, the spectrum is characterized by having a phase angle that increases indefinitely with angular velocity.

15. Regulating systems are ordinarily treated as having unlimited power supplies available. When this is *not* the case the power supply dynamics contribute to the total system behavior. If more than one regulator receives its power from the same supply then the regulators, rather than being independent systems, are coupled with the power supply to make one large system.

16. In addition to the normal regulating action exhibited by homeostats, many also have the ability to alter their properties under conditions of maintained stress or load. This is here termed an adjustment to distinguish it from adaptation, the latter designating changes that are evolutionary in character. The experimental study of homeostats can eliminate those changes due to adjustments by employing sinusoids as the stimuli.

17. Disturbances may appear at any point within a feedback system, but regardless of their location and character, the regulator should minimize their effect upon the regulated variable. In contrast, an adjustment in a sense compensates for disturbances and enhances the quality of the regulation.

18. The multivariable character of homeostats serves to increase the possible operating range as well as promote reliability. The multiplicity of receptors associated with such a multivariable system may be the means of providing an effective reference input whose identification in many homeostats is not clear. Homeostats are frequently arranged in a hierarchy, a fact that further complicates the experimental study of many systems. Physiological variables are further distinguished by invariably being nonnegative quantities.

Problems

1. From Fig. 8-1a, it appears that the body temperature of man stays quite constant even though the external temperature is below or above that of the body.

(a) Discuss the effectiveness with which various species (echidna, rodent, bird, dog, cat) are able to maintain a body temperature *below* that of the surroundings, and the mechanisms for accomplishing this (sweating, panting).

(b) In a sense, sweating serves as a heat pump. Describe the role of evaporation and ambient humidity on this process.

2. Figure 8-2 shows the operating regions for the three temperature-regulating mechanisms as functions of the external temperature. In a real sense, this figure is misleading and erroneous, even assuming that the linear relations shown are quite correct. Explain why this is so.

3. Inasmuch as all the individual thermoreceptors exhibit a peak

response, it is a possibility that these units might be called upon to operate on either side of the peak.

(a) What would be the effect insofar as the temperature regulation is concerned of operating at temperatures below the peak of the cool receptor characteristic, or above the peak for the warm units? See Fig. 8-3.

(b) What appears to be the effect of considering a population of units rather than a single one?

(c) How do the temperatures at which these peaks occur compare with the lethal temperatures?

4. (a) Are the core temperature receptors shown in Fig. 8-5 of the "warm" or "cool" variety?

(b) What about the core receptors in Fig. 8-6?

5. Thermoreceptors are described as tonic units having a steady discharge frequency that is proportional to temperature in the manner shown by the characteristic curves in Fig. 8-3. Is the tonic character of their behavior a necessary feature for temperature regulation, or simply a byproduct of their internal constitution? Explain.

6. Describe in general terms the changes in energy sources, in the flow patterns for thermal energy and material, and in the thermal loss mechanisms associated with each of the following control actions. Note that in each instance the control involves more than a simple valve action such as shown in the fluid flow models discussed previously.

(a) Shivering.

(b) Vasodilitation.

(c) Sweating.

7. (a) Estimate the ratio $\Delta T_r / \Delta T_h$ from the data of Fig. 8-10.

(b) Construct a simplified block diagram for the temperature regulating system corresponding to this figure to show both the circulatory and neural pathways to the hypothalamus. From this diagram justify the statement on p. 269 that the ratio computed in (a) is probably in error.

(c) Cite an additional reason why the ratio calculated in (a) cannot be identified with the loop gain.

8. From the discussion of Section 8.10, it appears that the effect of a given disturbance on a known system may be calculated. However, the converse, that of identifying a disturbance that produces a given effect, it is not soluble. In Fig. 8-9 assume that $K = 100$ and $K_1 = K_2 = 10$.

(a) Verify the expressions for X_4 in Fig. 8-9a and b.

(b) If ΔX_4 in Fig. 8-9a is observed to be 0.01, find the ΔX_1 producing it.

(c) If ΔX_4 in Fig. 8-9b is also 0.01 and X_1 is constant, find the ΔX_5 producing it. Note that an observed perturbation of X_4 may be caused by quite dissimilar disturbances at different points within the system.

(d)   If $\Delta X_5 = 0.01$, calculate ΔX_4. What conclusion can you draw from this?

9.   (a)   Compute a few points on the spectrum of $\exp[-(j\omega)^{1/2}]$ as shown in Fig. 8-13c. (Hint: $\exp[-(j\omega)^{1/2}] = \exp[(\omega/2)^{1/2} - j(\omega/2)^{1/2}]$ so that $|\exp[-(j\omega)^{1/2}]| = \exp(\omega/2)^{1/2}$, and $\angle \exp[-(j\omega)^{1/2}] = -(\omega/2)^{1/2}$.)

(b)   Estimate the critical gain for this feedback system.

10.   A number of the figures in this chapter depict experimental observations in which the independent variable is some measure of environmental temperature variously denoted as external, effective, skin, or calorimeter temperature. Discuss these several measures as regards (a) their relation to the body thermostat, and (b) as an accurate measure of the thermal environment.

References

Benzinger, T. H., and Taylor, G. W. (1963). Cranial measurement of internal temperature in man, *in* "Temperature, Its Measurement and Control in Science and Industry," Vol. 3, pp. 111–120. Van Nostrand-Reinhold, Princeton, New Jersey.

Copp, D. H. (1969). Endocrine control of calcium homeostasis. *J. Endocrinol* 43, 137–161.

Du Bois, E. F. (1941). The temperature of the human body in health and disease, *in* "Temperature, Its Measurement and Control in Science and Industry," Vol. 1, pp. 24–40. Van Nostrand-Reinhold, Princeton, New Jersey.

Eisenman, J. S., and Edinger, H. M. (1971). Neuronal thermosensitivity. *Science* 172, 1360–1362.

von Euler, C. (1950). Slow "Temperature potentials" in the hypothalamus. *J. Cell. Comp. Physiol.* 36, 333–350.

von Euler, C. (1961). Physiology and pharmacology of temperature regulation. *Pharm. Rev.* 13, 361–398.

von Euler, C. (1964). The gain of the hypothalamic temperature regulating mechanisms. *Progr. Brain Res.* 5, 127–131.

Fleming, D. G., Mesarovic, M. D., and Goodman, L. (1964). Multi-level multi-goal approach to living organisms, *in* "Neuere Ergebnisse der Kybernetik" (K. Steinbuch and S. W. Wagner, eds.), pp. 269–282. R. Oldenbourg KG, Munich.

Grodins, F. S., Gray, J. S., Schroeder, K. R., Norins, A. L., and Jones, R. W. (1954). Respiratory responses to CO_2 inhalation. A theoretical study of a nonlinear biological regulator. *J. Appl. Physiol.* 7, 283–308.

Green, D. E., and Goldberger, R. F. (1967). "Molecular Insights into the Living Process." Academic Press, New York.

Guyton, A. C. (1971). "Textbook of Medical Physiology," 4th ed. Saunders, Philadelphia, Pennsylvania.

Hammel, H. T. (1965). Neurons and temperature regulation, *in* "Physiological Controls and Regulations," (W. S. Yamamoto and J. R. Brobeck, eds.). Saunders, Philadelphia, Pennsylvania.

Hammel, H. T. (1968). Regulation of internal body temperature. *Annu. Rev. Physiol.* **30,** 641–710.

Hammel, H. T., Jackson, D. C., Stolwijk, J. A. J., Hardy, J. D., and Strømme, S. B. (1963). Temperature regulation by hypothalamic proportional control with an adjustable set point. *J. Appl. Physiol.* **18,** 1146–1154.

Hensel, H. (1963). Electrophysiology of thermosensitive nerve endings, *in* "Temperature, Its Measurement and Control in Science and Industry," Vol. 3, pp. 191–198. Van Nostrand-Reinhold, Princeton, New Jersey.

Hensel, H., and Boman, K. K. A. (1960). Afferent impulses in cutaneous sensory nerves in human subjects. *J. Neurophysiol.* **23,** 564–578.

Hensel, H., Iggo, A., and Witt, I. (1960). Quantitative study of sensitive cutaneous thermo-receptors with C afferent fibers. *J. Physiol.* **159,** 113–126.

Hensel, H., and Zotterman, Y. (1951). Quantitatis Beziehungen zwischen der Entladung einzelner Kaltfasern und der Temperatur. *Acta Physiol. Scand.* **23,** 291–319.

Jones, R. W. (1969). Biological control mechanisms, *in* "Biological Engineering." McGraw-Hill, New York.

Lind, A. R. (1963). Tolerable limits for prolonged and intermittent exposures to heat, *in* "Temperature, Its Measurement and Control in Science and Industry," Vol. 3, pp. 337–345. Van Nostrand-Reinhold, Princeton, New Jersey.

Mesarovic, M. D. (1968). Systems theory and biology — View of a theoretician, *in* "Systems Theory and Biology," pp. 59–87. Springer-Verlag, Berlin and New York.

Precht, H., Christophersen, J., and Hensel, H. (1955). "Temperatur und Leben." Springer-Verlag, Berlin and New York.

Prosser, C. L. (1964). Perspectives of adaptation; Theoretical aspects, *in* "Handbook of Physiology," Sect. 4, Adaptation to the Environment. Amer. Physiol. Soc., Washington, D.C.

Riggs, D. S. (1963). "Mathematical Approach to Physiological Problems." Williams & Wilkins, Baltimore, Maryland.

Stolwijk, J. A. J., and Hardy, J. D. (1966). Temperature regulation in man — A theoretical study. *Pfluegers Arch.* **291,** 129–162.

Stuart, D. G., Eldred, E., Hemingway, A., and Kawamura, Y. (1963). Neural regulation of the rhythm of shivering, *in* "Temperature, Its Measurement and Control in Science and Industry," Vol. 3, pp. 545–557. Van Nostrand-Reinhold, Princeton, New Jersey.

Wissler, E. H. (1964). A mathematical model of the human thermal system. *Bull. Math. Biophys.* **26,** 147–166.

Nonlinear Systems || *Chapter 9*

Although a linear system may be complex, having many components and interconnections, its dynamic behavior is relatively simple in the sense that it can be understood in terms of a few general principles. The dynamic response of even the most complex linear system can be expressed in terms of its modes of free vibration, and although there may be a very large number of such modes, they are of only two different types: decaying exponentials and damped sinusoids. In contrast, nonlinear systems may exhibit a greater variety of behavior patterns, many of which are far from obvious and not attainable with linear components. With even a single nonlinear component, it is no longer possible to apply the superposition principle, and the response must be calculated in some appropriate manner for each stimulus or disturbance, with few general principles to go upon.

In the previous discussions of linear systems, the effects of nonlinearities have been introduced at a number of points with the suggestion that linear approximations may be valid. In classical physics, this is termed a *small signal analysis,* or the *theory of small oscillations*. The essence of the linear approximation lies in replacing a nonlinear characteristic curve by a straight line tangent to it at the operating point. So long as the signals are small, and the line remains a good approximation to the actual curve, this method is very helpful. On this basis much of the linear theory in the previous chapters can be applied to the analysis of nonlinear systems. The real limitations to small signal analysis arise when the physiological signals can no longer be considered small, and when the nonlinear system exhibits patterns of behavior not attainable with linear components.

In this chapter we shall concentrate on the identification of nonlinear behavior, which must often be accomplished from an examination of the response pattern following a known stimulus, rather than from a consideration of the process itself. It will be found that to establish linearity, a number of tests must be examined, whereas nonlinearity may be evident from a single response. Furthermore, it will be of interest to consider those kinds of responses that are impossible to obtain from systems of linear components. Since the scope of nonlinear analysis is extremely broad, it will be necessary to restrict the discussion very largely to questions of relevance to homeostatic systems, and in particular to develop a physical picture of nonlinear behavior.

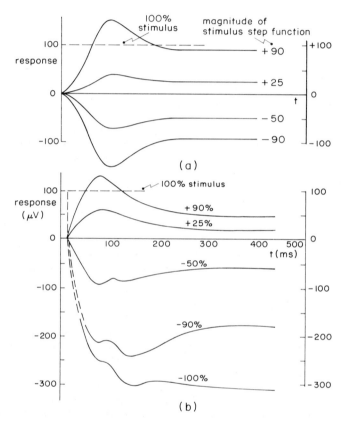

Fig. 9-1. (a) Responses of a linear system to step function input disturbances of the magnitudes indicated. (b) ERG responses of the wolf spider eye measured in microvolts for step function changes in illumination denoted as percent of ambient background illumination. From DeVoe, *J. Gen. Physiol.* **50,** 1967, 1961, by permission.

a. In order to identify the response characteristics of nonlinear systems, we will discuss some specific examples. In part (a) of the figure is sketched the response of a hypothetical linear system to a number of step function changes in the input (stimulus). The steps are identified simply as percentages of some maximum value, and are of both positive and negative sign.

b. For a linear system, the response curves for a given type of stimulus are all of the same shape, and are simply scaled in amplitude according to the size of the stimulus. Note that in (a), both the transient and steady-state portions of the curves change in direct proportion to the magnitude of the stimulus step. The shape of the response curve is also preserved for negative stimuli. Thus, all four of these response curves would superimpose upon an appropriate change in vertical scale and sign.

c. As the stimulus magnitude is increased, there will always appear a point at which the response amplitude no longer changes in a proportional manner. That is, there is a limit to the region of linear behavior and it should be adequately specified.

d. In contrast to the set of responses of a linear system, there is shown in (b) a similar set obtained from a system that is not linear. These particular curves were obtained from the wolf spider eye, for which the input was a change in background illumination, and the response was the ERG (electroretinal potential) measured across the eye, cornea to indifferent electrode. The departure from linearity is shown by a number of qualitative differences when compared to the curves in (a).

e. The shape of the curves changes markedly with the magnitude and sign of the stimulus. Comparing the two positive responses in (b), the percent overshoot and the final values do not bear a linear relation to the stimulus magnitude. Furthermore, the negative responses are qualitatively different from the positive ones, an effect not obtainable with linear components.

f. Note that for this transducer, represented by the curves in (b), the stimulus and response are quite different physical quantities (illumination, ERG). This is, of course, a common feature of all transducers and is not in any sense to be associated with the nonlinear behavior.

g. Although the responses shown in part (b) are clearly those of a nonlinear system, it might be possible to use the methods of small-signal analysis and design experiments to obtain the linear behavior for small stimuli. If this were done the responses would then be identified as those of a linear process and thus quite different from the ones shown in this figure. This matter is discussed further in the next section.

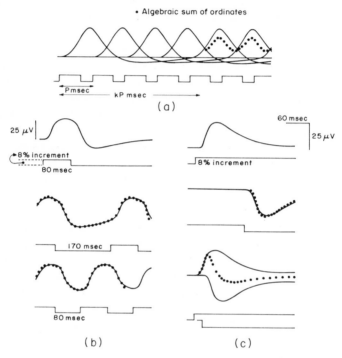

• Algebraic sum of ordinates

P msec

kP msec

(a)

25 μV

8% increment

80 msec

60 msec

25 μV

8% increment

170 msec

80 msec

(b)

(c)

Fig. 9-2. Application of the superposition principle to test the linearity of the wolf spider eye. From DeVoe, *J. Gen. Physiol.* **46,** 1962, 75, by permission. Stimulus is eye illumination and the response is the ERG. (a) Responses to a single pulse are superimposed to form the response to a pulse train (filled circles). (b) Flicker response predicted from the superposition of responses to a single pulse (filled circles). This agrees well with the experimental flicker response (solid line). (c) Superposition of the responses to a positive step (upper curve), and to a negative step (middle curve), to yield the pulse response. Filled circles in the middle curve are the negative of the response to a positive step.

a. The full significance of a linear system and the role of superposition in predicting the response to various stimuli can be best illustrated with a specific example. The eye of the wolf spider is again selected, because experiments are available which aptly demonstrate the consequences of linearity (DeVoe, 1962, 1963). The experimental procedures are the same as those described in the previous section, except that the magnitudes of the stimuli were small.

b. The linearity of the system is tested by using the principle of superposition in a number of ways. In (a), the response to a single pulse is shifted in time by an amount equal to the period of a pulse train, and the sequence of individual pulse responses is then summed to yield the steady-state response to the pulse train. The calculated pulse train response (circles) was found to be essentially the same waveform as was obtained by direct measurement; the latter is not shown in the figure.

c. The figure suggests that the waveform with a pulse train input is not strictly periodic during the first few pulses. This is the transient period, after which the response is truly periodic and of the same shape as that measured experimentally. The reader is referred to the original paper (DeVoe, 1962) for details regarding the variability of the data and accuracy with which the measured waveform can be predicted. Suffice it to say that the results justify the use of a linear approximation for small stimuli.

d. Similar results were obtained for a variety of pulse widths and stimulus periods; one example is shown in (b). Note the change in waveform with period, even though the same basic pulse width is used for both cases. While the response is periodic, it is not sinusoidal.

e. A rather impressive example of superposition is that of a positive and a negative step function response combined to yield a pulse response, as in (c). The responses to the two step functions in the steady state are equal but of opposite sign, and it is only during the pulse stimulus and immediately thereafter that there is a nonzero response.

f. As noted above, these are small signal experiments for which linearized equations appear to provide an entirely satisfactory representation. A portion of the linearizing procedure is that of replacing any curved characteristic by its tangent, and other types of substitution may be required for other forms of nonlinearity. Although the linear approximation is a very useful method in the study of many questions, it is not always applicable.

9.3 SUPERPOSITION OF RESPONSES   297

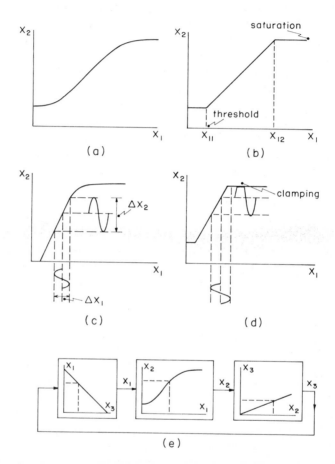

Fig. 9-3. (a) Steady-state characteristic curve exhibiting threshold and saturation effects with (b) its piecewise linear approximation. (c) Linear transmission for small signals, and (d) clamping effect of saturation with large signals. (e) Feedback system with nonlinear component.

9 NONLINEAR SYSTEMS

a. A common type of nonlinearity is that of a steady-state characteristic curve having a sigmoid or S-shaped form, such as depicted in (a). This curve is characterized by an intermediate portion that is relatively straight, but at both the low and high ends shows considerable departure from linearity.

b. At low values of the independent variable X_1, the curve exhibits a *threshold* below which the slope is very small, or even zero, so that no change appears in the output signal until the input exceeds the threshold value. In the curve in (c), there would be no output below threshold; a characteristic of many receptor organs.

c. *Saturation* occurs for large values of the input, and although all the curves in this figure show complete saturation, with the slope going to zero, varying degrees of saturation are also possible. Thus, both threshold and saturation effects will occur if the slope of the curve simply decreases at low and high values of X_1, without the slope actually falling to zero.

d. In the presence of both threshold and saturation, there is frequently an intermediate region in which the slope is quite constant and the relation between input and output variables is essentially linear. This would be the case in (b) if the variable X_1 never exceeded the range from X_{11} to X_{12}, and within this region the component can be treated as a linear one. It is of interest to note that in the case of muscle and other physiological components having this S-shaped characteristic, the normal operating region lies along the linear portion of the curve.

e. Parts (c) and (d) of the figure illustrate the consequences of operating within or outside of the linear region. In the case of complete saturation as in (d), the output signal never exceeds the saturated value and X_2 is said to be *clamped*. With saturation, no signal is propagated through the component. A similar situation occurs when the input falls below threshold.

f. A feedback system having a component of this character appears in (e) for which the operating point may be found in the usual manner. It will be clear that the operating point must be above threshold and below saturation, because otherwise there would be no signal transmission around the loop in the steady state.

g. If the sigmoid curve can be represented by three straight line segments, the system is said to be *piecewise linear,* and linear analysis may be applied in each region. An extreme case occurs when the slopes below threshold and above saturation are both zero, in which event the feedback loop is opened whenever the signal is driven into these regions.

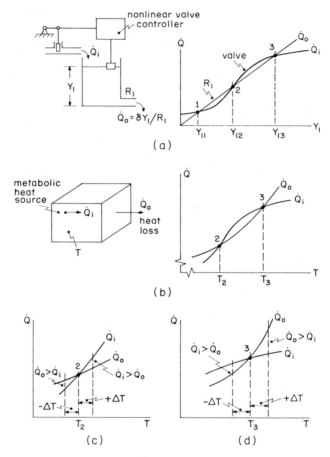

Fig. 9-4. Nonlinear steady-state characteristics leading to possible stable and unstable operating points. (a) Liquid level regulator with three possible operating points. (b) Heat flow from a region having a heat source. (c) Unstable operating point. (d) Stable operating point.

300   9   NONLINEAR SYSTEMS

a.   The previous discussion of stability suggested that disturbances to a system may result in oscillations that increase in amplitude with time until they result in failure of some component, or in maintained oscillations at an amplitude set by saturation.

b.   Another form of instability may arise in a process such as the one represented by the liquid level regulator in (a). The valve controller is presumed to have an S-shaped steady-state characteristic, while the outflow $\dot{Q}_o$ remains a linear function of Y_1.

c.   The presence of a nonlinear steady-state characteristic results in the possibility of multiple operating points; in this instance there are three, at which $\dot{Q}_i = \dot{Q}_o$. The question then arises, at which of these points does the system operate?

d.   A similar situation arises in (b), which depicts a compartment having metabolic processes generating heat at the rate $\dot{Q}_i$, with heat loss at the rate $\dot{Q}_o$. Both of these rates depend upon the temperature T in the manners shown. In the steady state, $\dot{Q}_i = \dot{Q}_o$, and there are two possible operating points.

e.   The relations existing at point 2 in both (a) and (b) are examined in more detail in (c). Assume that the system was operating at T_2, and a disturbance caused a momentary increase in temperature $(+\Delta T)$. As a consequence, $\dot{Q}_i$ would exceed $\dot{Q}_o$ and a further increase in T would follow. In a like manner, if T decreased $(-\Delta T)$, then $\dot{Q}_o > \dot{Q}_i$ and the temperature would fall further. Inasmuch as any disturbance at point 2 brings about an effect that is cumulative, the process is deemed unstable.

f.   A similar argument applied to operating points 1 and 3 will show that a disturbance $+\Delta T$ leads to $\dot{Q}_o > \dot{Q}_i$, and the process will return to its previous state when the disturbance disappears. Operation at Y_{13}, or T_3, is thus termed stable.

g.   These conceptual tests are based upon the following physical reasoning. (1) Disturbances are continually displacing the operating point in one direction or another. (2) The conditions set up by this displacement may either cause the operating point to move further away from its initial location (unstable) or tend to return it to its initial position (stable). The above relations provide a necessary, but not sufficient, condition for stability.

h.   A physical process having two stable states separated by an unstable one is termed *bistable,* and in many cases it may be "switched" from one stable state to the other upon receipt of an appropriate signal.

A common example is the ordinary wall switch which is either "off" or "on," but numerous other examples exist in electronics and a variety of physical processes.

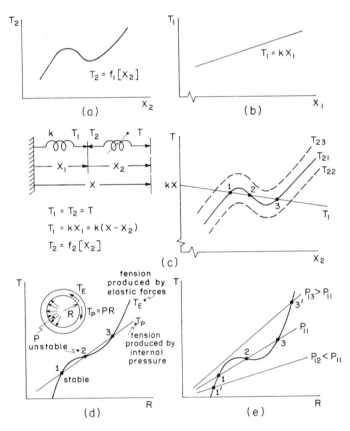

Fig. 9-5. Switching action in an elastic component with one nonlinear spring. (a) Tension-length diagram of a nonlinear spring. (b). Similar diagram for a linear elastic element. (c) Linear and nonlinear elastic elements in series with possible operating points 1, 2, 3. (d) Graphical construction for finding the tension in the wall of an arteriole. (e) Change in the radius of an arteriole with change in internal pressure under constant neural innervation (Burton, 1951).

a. The presence of multiple operating points, as suggested in the previous section, can lead to a form of behavior not possible with linear components. The S-shaped characteristic as in (a) may be observed in a nonlinear elastic element, and the existence of the inflection point can lead to switching action.

b. Assume, first, that this nonlinear spring (a) is attached in series with a linear elastic element (a simple spring) having the characteristic in (b). The two springs in series can exist in a steady state only if $T_1 = T_2 = T$, as shown in (c). Note that the linear spring is represented by a line of negative slope because the tension is plotted as a function of X_2; X is assumed to be held constant.

c. The construction in (c) shows three possible operating points, of which 1 and 3 are stable, whereas 2 is unstable. If the characteristic of the nonlinear spring is moved vertically a sufficient amount, then only one operating point exists and it is stable. A similar effect is produced if the characteristic of the linear spring is similarly displaced. Thus, a change in either of the two characteristics can cause the operating point to switch from one stable position to another, and its location will depend upon the previous history of the system. Relations similar to the above can be found in connection with muscles.

d. Consider an arteriole surrounded by smooth sphinctor muscle that serves to control the blood flow into a capillary bed. With an internal blood pressure P, the tension in the thin arteriole wall due to that pressure is $T_P = PR$. On the other hand, the elastic properties of the wall produce a nonlinear characteristic, such as shown in (d). These two characteristic curves may have a total of three intersections (Burton, 1951; Nichols *et al.*, 1951).

e. With a change in the internal pressure from P_{11} to P_{12} or P_{13}, there will be a corresponding change in the operating point, and in the radius R. Point 2 is unstable, so that continued operation at this point is impossible and the operating point will switch to 1 or 3 following some small disturbance.

f. Switching from one radius to another may occur as a result of a disturbance to either P or the muscular innervation. This would seem to account for the observation that capillaries appear to be either open or shut, and that graded control of blood flow at the capillary level does not occur.

g. A somewhat similar situation is believed to exist in the alveoli, which are observed to open and close in almost a switching mode. Although the individual arterioles or alveoli may act as two-position devices, the gross behavior of thousands of such units acting more or less in parallel may appear to be smooth or graded (Clements and Tierney, 1965).

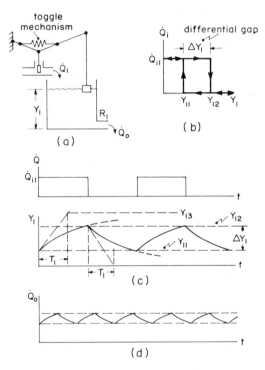

Fig. 9-6. (a) Liquid level regulator with a valve operated by a toggle mechanism arranged to produce a two-position controller; the valve is either completely open or shut. (b) Path traced by $\dot{Q}_i$ as Y_1 is caused to increase and decrease. $\dot{Q}_i$ is either $\dot{Q}_{i1}$ or 0, depending upon the magnitude of Y_1 and the past history, (c) the system cycles continuously with the magnitude of Y_1 oscillating between Y_{11} and Y_{12}. (d) The outflow $\dot{Q}_o$ oscillates in an identical manner. However, note that different time scales are used in (c) and (d).

9   NONLINEAR SYSTEMS

a. Having demonstrated in the previous section that switching action is not impossible in a physiological system, we now turn to the application of such a process to a feedback regulating system. Two-position regulators are found in a wide variety of technological applications; possibly the most frequently encountered examples are in temperature regulation in which a thermostat serves to switch a heat supply.

b. The valve control mechanism in (a) is the equivalent of the thermostat, but is a mechanical device to open and close the inlet valve. The levers and spring making up the control mechanism are so arranged that the valve can be either open or shut, and no other configuration is possible. Mechanically it is very similar to the wall switch controlling electric lights. Even though the liquid level rises very slowly, the valve itself remains open until a threshold value Y_{12} is reached, at which point the valve suddenly switches to its fully closed position.

c. This behavior is depicted in more detail in (b), where two operating thresholds are shown. If the valve is closed and the level falls to Y_{11}, the valve suddenly opens and $\dot{Q}_i$ jumps from zero to some fixed value. Later, as Y_1 increases to Y_{12}, the valve suddenly closes. In physical devices having this characteristic, there is a clear separation between Y_{11} and Y_{12}; this is termed the *differential gap,* or *hysteresis,* and refers to the fact that the state of the valve depends not only upon Y_1 but also upon the past history of Y_1.

d. The operation of this single-compartment regulator may be understood from (c), where the level is seen to rise and fall in an exponential manner. The level Y_{13} is the one that would be reached if the switching action did not intervene.

e. An important characteristic of two-position regulators is their continual oscillation, and while this mode of behavior was previously termed unstable, it becomes the normal mode here and one that may be quite acceptable. The question of acceptibility is largely concerned with the size of ΔY_1; in the case of the household thermostat the matter of a degree or so is of little significance. The differential gap may be reduced by making the switching mechanism more sensitive.

f. This system is piecewise linear, since its performance is entirely linear for each position of the valve. The concept of piecewise linearity can be extended to processes which operate linearly over not just two, but a number of limited regions, in each of which the parameters remain constant. As the variables move from one region to another the parameters take on different, but constant, values.

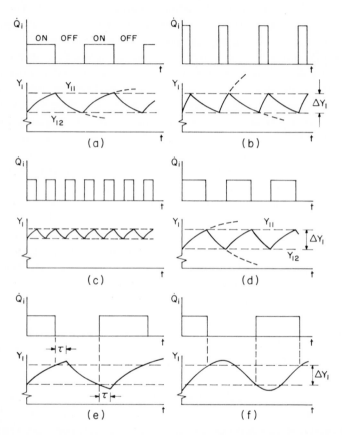

Fig. 9-7. (a) Behavior of the first-order system of Fig. 9-6 with equal ON and OFF times. (b) $\dot{Q}_i$ is increased so that the ON–OFF ratio is decreased. (c) The differential gap ΔY_1 of the measuring device is decreased, with a resultant increase in the operating frequency. (d) A decrease in the outflow resistance R_1 results in an increase in the ON–OFF ratio. (e) Introduction of a transport lag τ in the control pathway causes Y_1 to exceed the differential gap. (f) Multiple time constant system has the same effect.

9 NONLINEAR SYSTEMS

a. It has been suggested that both the shivering mechanism, as well as panting in response to a heat stress, are controlled by a two-position mechanism having some of the characteristics described in the previous section (Hardy and Hammel, 1963). It is possible that the peripheral circulation is also controlled by a similar mechanism.

b. In a two-position regulator, the final values of $\dot{Q}_i$ called for in each of the two valve positions must lie beyond the differential gap. That is, the switching must always occur before reaching the final value, and the more widely separated these two values are from the differential gap ΔY_1, the faster will be the response of the system and the higher its operating frequency. In (a), the final values of Y_1 are equally spaced on either side of ΔY_1, and the ON–OFF ratio is unity.

c. The effects of various disturbances are illustrated in (b)–(d). A decrease in outflow resistance R_1 is considered in (d). The differential gap remains the same. However, because R_1 has been reduced, the tank empties more quickly, and by the same token will take longer to fill. The exponential response curves have a time constant less than before, but the portions of the curve between the switching points Y_{11} and Y_{12} are different for the ON and OFF periods. The ON time is now greater than the OFF time, because the final values of Y_1 are no longer symmetrically displaced with respect to ΔY_1.

d. The preceding discussion was based upon a single-compartment process, and with the introduction of a multicompartment process the relations among the variables become quite different. The general effect may be seen in (f). Because of the presence of several compartments between the point of measurement and the point of control, there will be a delay in the system response, which shows up in this figure as an overshoot and an undershoot in the value of Y_1. This occurs because the corrective action must progress through several compartments before it can be sensed by the float, and as a consequence Y_1 may continue to rise or fall after the valve has already operated. The excursions in Y_1 now exceed the differential gap by an amount that depends upon the delays in the system. As a matter of fact, with a very large number of compartments, the oscillation approaches a sinusoid, and the analysis may sometimes be made on that basis.

e. The example described above employed a switching device in which the two positions were ON and OFF. Another possibility would be that in which the control is between two values HIGH and LOW. The general behavior is essentially the same.

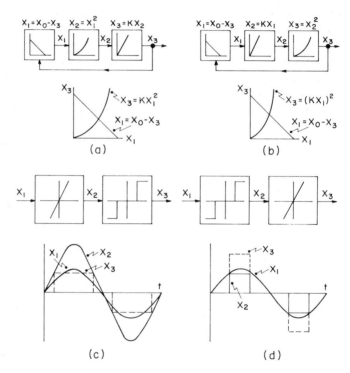

Fig. 9-8. Sequences of linear and nonlinear processes. (a) Feedback system with a nonlinear component preceding a linear one. (b) The same system, but with the two processes interchanged, has a different operating point. (c) Nonlinear component having both threshold and saturation follows a linear component. (d) The same processes as in (c) are interchanged, resulting in a different output signal for the same sinusoidal input.

a. In our previous discussions of linear systems, a point was made of the fact that the behavior was largely independent of the sequence in which the several processes appeared in the system. This was particularly true of a system of noninteracting components, for which we showed that the system behavior was completely independent of the sequence of the components, and use was made of this fact in the construction of frequency spectra. Interacting components produce some modification of this principle. With the introduction of nonlinear components, this matter of sequence of component processes around the loop will in general have a major influence on system performance. A few examples will demonstrate.

b. In (a) is a feedback system containing one nonlinearity $(X_2 = X_1^2)$, and in (b) the same components appear, but in a different sequence. The total steady-state transmission from X_1 to X_3 is quite different in the two cases, and this accounts for the change in operating point shown. One would also expect to find a difference in dynamic behavior as a consequence of this interchange of linear and nonlinear processes.

c. Parts (c) and (d) of the figure show the dynamic performance of two systems each containing a saturating component, and differing only in the location of that nonlinearity. For simplicity, no feedback is shown and the transmission of a sinusoidal signal is considered. The figures rather clearly show the difference in X_3 that results from the saturation preceding the amplifier, on the one hand, and following it on the other.

d. These examples indicate that the presence of a specific nonlinearity is not the only factor affecting system performance; its location in the feedback loop may also be significant. Thus, it is more difficult to generalize than with strictly linear systems, and one finds that each nonlinear system exhibits a performance that is largely unique to that system.

e. The above comments are applicable to feedback and nonfeedback systems alike. From the standpoint of model building, of the construction of a set of equations to represent the prototype, this situation suggests prudence in the identification of individual processes and the preservation of isomorphism between prototype and model.

f. A factor which tends to temper the above conclusions is the low-pass nature of most regulating systems. In general, components severely attenuate high-frequency components in the signal. The effect of this in some instances will be to make the matter of process sequence of somewhat less significance.

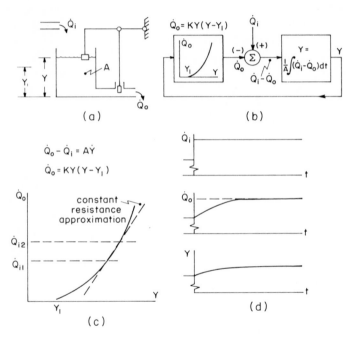

Fig. 9-9. (a) Liquid level regulator with control of the outflow resistance (rather than the inflow rate) makes this a variable parameter system. (b) Block diagram showing the sign reversal as an inherent part of the process. (c) Steady-state characteristic on which the operating points for two values of $\dot{Q}_i$ are indicated. (d) Behavior following a step function increase in $\dot{Q}_i$.

a. In the case of many biological processes, the nonlinear characteristic curve arises because what has previously been termed a constant parameter is in fact controlled by one of the dependent variables. An example is shown in (a), where the valve provides a variable outflow resistance, the magnitude of which varies inversely with Y. Such processes are also termed *facultative* (Yamamoto, 1965).

b. This process and its block diagram in (b) are similar to those shown in Fig. 4-12. However, the discussion of the earlier figure made no mention of the possible nonlinear behavior of the system. The characteristic for the control mechanism (float, linkage, valve) is based upon the assumption that $\dot{Q}_o$ depends upon both the head Y, as well as the position of the valve $K(Y - Y_1)$. The product of these two quantities yields a term Y^2, and results in the nonlinear characteristic in (c).

c. The operating points for two values of the inflow, $\dot{Q}_{i1}$ and $\dot{Q}_{i2}$, are shown in (c). Note that neither of the two steady-state curves has a negative slope. This results from the fact that the sign reversal required for negative feedback is inherent in the system, because the control is exercised over the loss function, $\dot{Q}_o$.

d. The dynamic behavior following a large step function increase in $\dot{Q}_i$ is sketched in (d). Although these curves may appear to be the same as those of a linear system (Fig. 5-1) the similarity is illusory. In the first place, the curves are *not* exponential, and therefore cannot be described by a time constant. Second, $\dot{Q}_o$ and Y are no longer linearly related and have different shapes.

e. The "resistance" to outflow becomes a variable parameter. However, for small disturbances the actual characteristic can be approximated by a straight line as indicated in (c), the latter being the characteristic for a constant resistance. For the disturbance considered in (c), the straight line is a good approximation to the actual curve. To the extent that such approximations can be made, the process may be represented by a linear model, and a small-signal analysis is possible.

f. A large number of biological processes are of the kind described; that is, they include parametric control, and the describing equations contain the product of two dependent variables (or a squared term). Although a good approximation to the system behavior may be obtained from a small signal analysis, a more complete study would require the aid of a computer. The behavior in (d) exhibits no qualitative differences from what one might expect with a linear model, but the similarity may become more remote as the disturbance becomes larger or the system complexity increases.

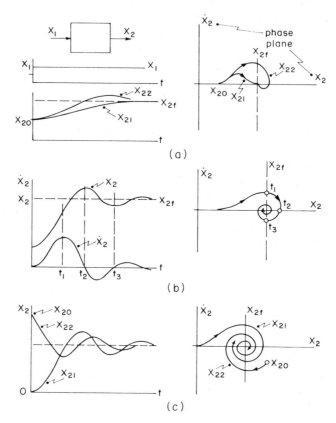

Fig. 9-10. Time response curves and corresponding phase plane trajectories for a linear system. (a) Response to a step function input of an overdamped system (X_{21}), and a slightly underdamped system (X_{22}). (b) Oscillatory response following zero initial velocity. (c) Oscillatory response following zero initial velocity (X_{21}), and negative initial velocity (X_{22}).

9   NONLINEAR SYSTEMS

a. Up to this point the dynamic behavior of a given system has been represented in either the time or the frequency domain. That is, the response to a given disturbance has been represented as a function of time much as it would appear on the face of an oscilloscope, or somewhat indirectly in terms of the sinusoidal behavior and its frequency spectrum.

b. Another mode of representation, termed the *phase plane* and mentioned briefly in Section 5.5, is frequently found useful. A number of examples are sketched in this figure. These represent the responses of a second-order linear system, without feedback, to a specified disturbance and initial conditions. The phase plane has the coordinates X_2, $\dot{X}_2$, and the trajectory on this plane depicts the displacement X_2 and the velocity $\dot{X}_2$ as they change with time. In (a), the response X_{21} starts at the initial conditions $X_2 = X_{20}$, $\dot{X}_2 = 0$, and proceeds to its final value $X_2 = X_{2f}$ without overshoot. With different system parameters, the response might exhibit one overshoot, as does X_{22}. The trajectory for X_{22} shows a portion with positive velocity, as well as one with negative velocity.

c. In (b) the response is a damped sinusoid with several overshoots, and the trajectory spirals around the final value before settling down at X_{2f}. Time is a variable that runs along the trajectory, and each point on the trajectory can be marked with elapsed time. The time t_1 corresponds to the instant of maximum $\dot{X}_2$, and t_2 to that of zero velocity. Note that when $\dot{X}_2$ is positive, X_2 is increasing, whereas when $\dot{X}_2$ is negative, X_2 is decreasing. Although theoretically the oscillation persists indefinitely, practically it approaches so close to the point X_{2f}, 0 that the oscillations are no longer observable.

d. Inasmuch as each point in the plane could be a set of initial conditions, each point could serve as the starting point for a trajectory. Two such cases are shown in (c). The phase plane is thus completely filled with trajectories, but only representative ones or ones of specific interest are actually drawn.

e. Some fundamental properties of the phase plane trajectories are readily stated. Through each point in the phase plane, it is possible to sketch a trajectory as indicated in the previous paragraph. A second property is the fact that through any one point only one trajectory may pass. If it were otherwise, then starting at that point, two different behaviors of the system would be equally possible. That is, the system would no longer be determinate but would exhibit a random behavior. This is not a possibility for the systems under discussion.

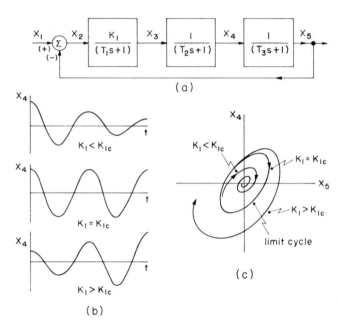

Fig. 9-11. (a) Linear feedback system of the third order, with a loop gain K_1. (b) Oscillatory responses with K_1 less than, equal to, and greater than its critical value K_{1c}. (c) Phase plane with coordinates X_5, X_4, showing the limit cycle for critical gain, and trajectories departing from the limit cycle for loop gains other than the critical value.

9 NONLINEAR SYSTEMS

a. At this point we wish to recapitulate some previous conclusions and relate them to the present discussion of nonlinear behavior. In particular, the criteria for instability of a linear system are reexamined, and the conditions necessary for continuing oscillation at a constant amplitude considered in more detail.

b. The third-order system shown in (a) may exhibit three different kinds of behavior as shown in (b), depending upon the relative values of loop gain K_1, and critical gain K_{1c}. Although the case $K_1 > K_{1c}$ is generally considered to be unstable, that in which $K_1 = K_{1c}$ (and the oscillations appear to persist indefinitely) might be quite acceptable in some applications, as suggested in Section 9.7.

c. The three possible response modes shown in (b) have been replotted on the phase plane in (c). This particular plane, having the coordinates X_4, X_5, is only one of several that could be selected, and while the detailed shape of the curves would be different on different planes, their general behavior is the same. In (c) it will be noted that for $K_1 < K_{1c}$, the trajectory approaches the origin with oscillations of decreasing amplitude, whereas for $K_1 > K_{1c}$ the amplitude of X_4 (and X_5) increases with each period.

d. For the case of critical gain, $K_1 = K_{1c}$, the trajectory forms a closed path termed a *limit cycle,* which although of an elliptical shape in this figure, may well take other forms. This depends upon both system equations and parameters, as well as the variables selected for the phase plane. It is thus seen that an oscillator may be described by its limit cycle, and the point representing the oscillatory state of the system follows this path indefinitely.

e. Returning to the linear system in (a), the case of critical gain is seen to require a precise relationship among all the system parameters. Assuming that is attained and a limit cycle ensues, subsequent disturbances are likely to upset this relation among the parameters, thus making continued oscillation around the limit cycle improbable.

f. From a more detailed examination of this "linear" limit cycle, it is found that a stable oscillator cannot be formed from linear components alone. The concept of stability must be widened to include that of *orbital stability,* defined as the maintenance of *amplitude and frequency* in the face of normal transitory disturbances. Stable oscillators contain some form of nonlinearity to give them the necessary orbital stability.

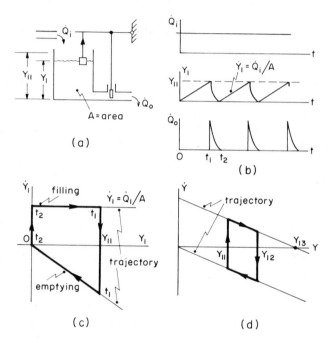

Fig. 9-12. (a) Tank and valve arranged to continually fill and empty the tank in the manner shown in (b). This is termed a relaxation oscillation and can hardly be considered a regulatory process. The filling (charging) phase normally occupies a much longer time interval than the emptying (discharge) phase. In (c), this periodic behavior is depicted on the phase plane $(Y_1, \dot{Y}_1)$ as a closed path, that is, a limit cycle. In this figure, the switching at Y_{11} and 0 is assumed to occur instantaneously, and is thus represented by the vertical lines. (d) Limit cycle shown for the two-position regulator of Fig. 9-6. In both (c) and (d) there are only two trajectories, because these are first-order systems.

a. The oscillations described in the previous section were sinusoidal in shape and were shown to arise from instability in a linear feedback system. The oscillations which might be produced by a system containing nonlinear processes would probably not be sinusoidal, but would still be a continuous curve of a periodic nature. A different form of oscillation appears, however, when the system contains a process having some form of switching that operates at a well-defined threshold. Such a system is termed a *relaxation oscillator*.

b. The system shown in (a) is distinguished by having a constant input flow $\dot{Q}_i$ and a control mechanism operating the outlet valve that functions only when the level has risen to the threshold value Y_{11}. At that point the valve, which was previously closed, opens completely (there is no proportional action) and the valve remains open until the tank has essentially emptied. When that state is reached, the valve closes completely and the tank starts to refill. It will be observed that this model bears certain resemblance (as well as significant differences) to a well-known household device.

c. With the valve having the characteristics enumerated above, the time behavior of the system will appear as in (b). With a constant input, the level rises at a constant rate until the threshold Y_{11} is reached. The valve is then "switched" to its open position and the tank empties along an exponential curve in a very short time. In fact, with most technological oscillators, the discharge time is negligible compared to the charging time, so that the frequency is largely set by the input rate and the tank area or capacity. The production of pulse trains by a pacemaker neuron somewhat resembles a relaxation oscillator.

d. Limit cycles for the relaxation oscillator, and the two-position regulator, appear in (c) and (d). The trajectories in these figures are straight lines, but this is distinctly a special case and a consequence of the first-order process used in both examples. The first-order differential equation, which completely defines the process, provides a relation between Y and $\dot{Y}$ so that these two quantities cannot be chosen as independent initial conditions. A first-order system thus has only a single trajectory in the phase plane, and it is a straight line. In contrast with the higher order systems, the plane is not filled with trajectories.

e. In a general sense, an oscillator converts a constant flow of energy (or material) into a periodic flow. In the examples discussed, this was accomplished by suitable phase relations in a continuous system (Fig. 9-11) or by a switching component in a discontinuous manner (Fig. 9-12).

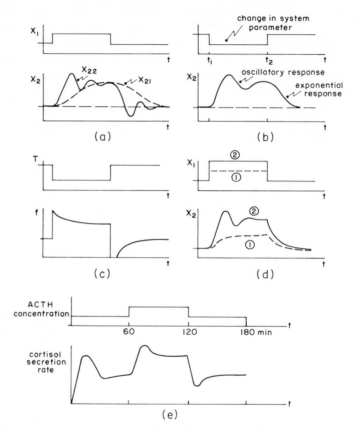

Fig. 9-13. (a) The disturbance (stimulus) X_1 impressed upon a linear process may yield either an exponential response X_{21}, or one that is oscillatory, X_{22}. In both cases the response to a positive step is of the same form as that to a negative step. (b) The disturbance produced by a parametric change in a linear system may result in asymmetrical responses. (c) Neural pulse frequency f of a cool temperature receptor in response to temperature changes. (d) Generator potential X_2 of a Limulus photoreceptor changes form as the magnitude of the step function change in light intensity X_1 is altered. (e) Behavior of the adrenal gland to changes in the ACTH perfusion rate.

9 NONLINEAR SYSTEMS

a. In the case of numerous receptors and other biological processes, the responses to stimuli frequently exhibit an asymmetry that is not expected of a purely linear component. However, asymmetry is possible even with completely linear processes, so that some further examination of this behavior is indicated.

b. If an input is suddenly applied to a series of compartments, the transient response may be either exponential or oscillatory, as in (a). If the system is linear, then it is generally expected that the form of the transient will be the same for application and removal of the disturbance, providing only that there are no threshold or saturation effects.

c. However, if the stimulus (disturbance) takes the form of a change in one of the system parameters, then the dynamic behavior may well be affected as was shown earlier in Fig. 3-5. Thus, a parametric disturbance may change a transient from exponential to oscillatory form or vice versa as suggested by (b), where the ON transient is oscillatory and the OFF transient exponential. As long as the disturbance is a step function, the parameter remains constant and the system is linear, albeit with different properties prior to and after the disturbance. The fact of asymmetry is by no means clear evidence for nonlinearity, but rather for a change in the parameters.

d. The cool temperature receptor shown in (c) may appear to be nonlinear, but only because the neural pulse frequency falls below threshold (zero) upon a return to the original temperature (Hensel, 1963). Although these units exhibit a rise in pulse frequency with a fall in temperature, this sign reversal should not be confused with nonlinear behavior. However, since negative pulse frequency is impossible, firing ceases at zero frequency.

e. A clearer example of nonlinear behavior is the Limulus photoreceptor in (d), where the character of the transient changes from overdamped to underdamped as the magnitude of the stimulus increases (Fuortes, 1958). A somewhat similar change appeared in Fig. 9-1.

f. Another example is that of the adrenal gland (e), the cortisol secretion rate of which is a function of the ACTH concentration with which it is perfused (Urquhart, Krall, and Li, 1968). The overshoot in secretion rate changes markedly with the concentration of ACTH, and there is asymmetry between the ON and OFF responses (Urguhart and Li, 1968).

g. The subject of asymmetry in response characteristic, or *unidirectional rate sensitivity* as it is sometimes called, has been examined in more detail in a recent volume (Clynes, 1969).

In the study of any physiological system, it should be assumed at the outset that it contains some nonlinear processes, and that the possibility of finding linear relations must be established by experiment. Fortunately, many systems can be adequately described by linear equations, but only over limited regions, and one task of the experimenter is to establish these regions in which small-signal analysis can yield acceptable results. For this purpose the superposition principle is an indispensable tool.

Figure 9-10c suggests that a linear system, having a single steady-state operating point, will attain that point from any set of initial conditions consistent with the equations of the system and the physical constraints. In contrast, the nonlinear system, frequently having more than one final operating point (stable or unstable), will follow a trajectory that may be quite different for various choices of initial conditions. That is, the choice of one set of initial conditions over another may lead to very different consequences insofar as the subsequent behavior is concerned.

This situation can lead to some significant properties of the system. In changing the initial conditions by only a small amount, it is possible to drastically alter the behavior. In an extreme but perfectly possible case, a small change in the initial conditions may cause the system to move toward a stable operating point on the one hand, and an unstable one on the other. A dramatic instance of this phenomenon is provided by the competition between animal species as described by the Volterra equations (Cunningham, 1958). Although initially the number of individuals in the two species may be quite small, a small difference in their relative numbers can result in entirely different final states. From the standpoint of experimental protocol, the necessity of maintaining a constant set of initial conditions for successive applications of a given stimulus will be obvious.

In the above discussion, it was tacitly assumed that any set of initial conditions might prevail in a homeostatic system. In the actual operation of a homeostat, is it possible that the initial conditions vary over as wide a field as this statement implies? This can be answered by noting that most systems are being disturbed continually, and that the system no sooner starts on one course than it is beset by another disturbance that again changes its path or trajectory. From this viewpoint, it would appear that any set of values for the several variables that is physically attainable may become the initial conditions for its subsequent behavior.

Some idea of the way in which initial conditions may affect the subsequent motion of a system is well illustrated by Fig. 10-2c. Although this figure is not a phase plane, it is clear that each of the four curves starts with different initial conditions. The initial magnitude $(a - a_\infty)$ is the same for all curves, but their initial velocities are quite different. If the trajectories were plotted on the phase plane, each one would start from a different point, but they would all approach the origin with infinite time (Problem 9-8). This attribute of linear systems, the fact that the steady state is independent of the initial conditions, has been termed *equafinality*, a property which nonlinear systems may not have.

In the brief discussion of the phase plane on the previous pages, the examples were all of the first or second order. These systems take no more than two variables to describe their state, so that all trajectories can be drawn on a single plane. Passing to the nth-order system, it is found that n variables are required to define the state, and therefore trajectories will occupy a space of n dimensions. The complexities of such a graphical representation in so-called *phase space* are obvious, but many of the concepts developed for second-order systems can be carried over for systems of higher order.

In the example of the Volterra equations, it is possible to draw a curve on the phase plane that separates those initial conditions that lead to one final value from those that lead to a second final value. This line is termed the *separatrix* because it separates all initial conditions into two sets, depending upon the terminal point of the trajectory. In a higher-order system, the separatrix becomes a surface, and serves the same function. This is but one example of the way in which the phase plane representation provides a broad picture of system behavior under a variety of conditions, even though it does not directly provide a detailed picture of the response in a specific case.

Epitome 9.16

1. Although a system containing nonlinear components may exhibit modes of response that are impossible to obtain with a strictly linear system, it is frequently possible to restrict the analysis to the effects of small disturbances for which the system behaves in a linear manner. When this is done, the aspects of behavior that are the unique consequences of the nonlinearities are of course lost.

2. The qualitative character of the response of a linear system is not affected by changes in the magnitude or sign of the disturbance. In contrast, with nonlinear components, changes in the intensity or sign of the stimulus may markedly affect the character of the response.

3. A sensitive test for linearity is provided by application of the superposition principle to experimental data. The successful use of pulse and step function responses in the prediction of other responses provides reasonably satisfactory evidence that the system may be considered linear for other disturbances.

4. An S-shaped characteristic curve, commonly found for a variety of biological processes, is characterized by both a threshold value and saturation. It is frequently found that the normal operating range occurs over the center portion of this curve, where the relations are close to linear.

5. Steady-state characteristics that are curves invariably lead to the possibility of several operating points whose stability must be evaluated. The stable and unstable points interlace each other, and the point at which the system actually operates depends upon its past history.

6. With two stable operating points, it becomes possible to switch from one to the other by appropriate displacement of one of the characteristic curves. By this means, a two-position component is obtained whose operation may be switched from one state to another upon receipt of a suitable signal.

7. A feedback controller employing a two-position device results in a regulating system that is continually oscillating about some desired value. The oscillations may be quite acceptable in some applications; their magnitude and frequency can be altered by suitable design.

8. A two-position regulator adjusts to system disturbances by changing the ratio of ON to OFF times. For a single time-constant system, the excursions in the regulated variable are equal to the differential gap, whereas with time lag or a higher-order system, the regulated variable will reach values outside the differential gap.

9. In contrast to a linear system, the sequence in which the components in a nonlinear system are arranged does affect the signal transmission and therefore the dynamic behavior. One concludes from this fact that isomorphic relations must be strictly maintained in model construction.

10. Many biological regulators use a controllable parameter to produce regulation. This results in a variable-parameter (or facultative) system, the equations of which, although nonlinear, can frequently be approximated by a linear model.

11. The dynamic behavior of physical systems can be represented by trajectories on a phase plane along which moves the representative point whose location at any instant describes the state of the system. Time becomes a distance along the trajectory.

12. Although a linear system with critical gain will theoretically oscillate indefinitely, this does not prove to be a stable oscillator. An oscillator is depicted on the phase plane by a closed trajectory, the limit

9 NONLINEAR SYSTEMS

cycle. If all trajectories in the neighborhood of the limit cycle approach it with increasing time, the oscillator is deemed stable and able to maintain its amplitude and frequency.

13. The combination of a storage process, and a means for discharging it when the quantity stored reaches some threshold value, produces a relaxation oscillator. The operating frequency is normally set by the time required to charge the storage volume.

14. Asymmetrical responses, although frequently a sign of non-linearities, are not conclusive evidence for their presence. Better evidence is a qualitative change in the response with a change in the stimulus amplitude, and the failure of superposition.

15. The important role played by the initial conditions in the behavior of a nonlinear system is clearly revealed by the trajectories drawn in phase space. Although these trajectories can provide all information necessary for the temporal response to a given disturbance, it is in the global picture they reveal of the general features of system behavior that their value lies.

Problems

1. The transient portions of the negative responses in Fig. 9-1b exhibit two properties not found in linear systems, and clearly identify the system as nonlinear.
    (a) What are these two properties?
    (b) The fact that the curves in Fig. 9-1a have an initial rate of change that is zero, whereas those in (b) are nonzero, has nothing to do with the nonlinearity. To what might this difference in initial conditions be attributed?

2. Using the data in Fig. 9-2a, sketch the response beginning at the first pulse. Does the shape of the individual pulse response give some clue to the duration of the initial transient?

3. The steady-state characteristic of skeletal muscle (force vs. neural pulse frequency) has a shape similar to Figs. 9-3a and 2-11d. Actual curves may be found in the literature (Cooper and Eccles, 1930). Find the maximum gain for the following muscles:
    (a) Soleus.
    (b) Internal rectus.
    (c) Gastrocnemius.
    (d) Extensor digitorum longus.

4. Assume that the variables in Fig. 9-4a are related by the following

three expressions (rather than the characteristic curves shown in that figure):

$$\dot{Q}_i = a(Y_1 - Y_o), \qquad \dot{Q}_o = bY_1, \qquad \text{and} \qquad A\dot{Y}_1 = \dot{Q}_i - \dot{Q}_o.$$

(a) Show that if $b < a$, the system is unstable, whereas it is stable if $b > a$, and $Y_0 < 0$. (Hint: use the differential equation.)

(b) Does this result correspond with the discussion in Section 9.5?

5. Figure 9.5e shows possible relations between $T_P = PR$ and $T_E = f_2[R]$.

(a) As P is increased from P_{12} to P_{13}, trace transition from $1'$ to $3'$. In particular, find the values of P_1 at which the transition from one value of R to another takes place.

(b) Same as (a), but let $T_E = f_2[R]$ change as a consequence of neural innervation. Assume this characteristic is simply translated vertically with change in neural innervation.

6. Make a sketch similar to Fig. 9-6, but in which the switching points Y_{11} and Y_{12} are moved up or down keeping ΔY_1 constant. What effect does this have on the relative OFF and ON times?

7. For the two systems shown in Fig. 9-8a and b, let $K = 2$. Find the operating point for each of these two configurations.

8. Make a rough sketch on the phase plane, the coordinates of which are $(a - a_\infty)$, and $d(a - a_\infty)/dt$, of the four response curves shown in Fig. 10-2c. Pay particular attention to the different initial conditions.

9. Consider the liquid level regulator of Fig. 9-9.

(a) If the horizontal characteristic curve $\dot{Q}_o = \dot{Q}_i$ could be given a small positive or a small negative slope, would the operating point be unstable or stable? (Hint: use the method of Fig. 9-4.)

(b) If $\dot{Q}_i$ were to be suddenly increased from $\dot{Q}_{i1}$ to $\dot{Q}_{i2}$, in Fig. 9-9, indicate the path (trajectory) followed during the ensuing transient.

(c) Same as (b), but for a sudden increase in K.

10. (a) In Fig. 9-12c, show that equal intervals along the horizontal segment $t_2 - t_1$ represent equal time intervals.

(b) Is this also true for the slanting trajectory $t_1 - t_2$? Explain.

(c) Develop the differential equations that apply to the system in this figure. Note that different equations apply to different portions of the operating cycle.

11. Contrast the behavior of a dependent variable (the output) in a physical process (without feedback) for each of the cases given below. A qualitative answer will suffice.

(a) Linear process with a single sinusoidal input.

(b) Nonlinear process with a single sinusoidal input.

9   NONLINEAR SYSTEMS

(c) Linear process with an input consisting of two sinusoids of different frequencies.

(d) Nonlinear process with the same input as in (c).

(e) Would the answers in any of the above be qualitatively different if the process had been a feedback system?

References

Burton, A. C. (1951). On the physical equilibrium of small blood vessels. *Amer. J. Physiol.* **164,** 319–329.

Clements, J. A., and Tierney, D. F. (1965). Alveolar instability associated with altered surface tension, *in* "Handbook of Physiology," Sect. 3, Vol. 2, pp. 1565–1583. Amer. Physiol. Soc., Washington, D.C.

Cooper, S., and Eccles, J. C. (1930). The isometric responses of mammalian muscles. *J. Physiol.* **69,** 377–385.

Clynes, M., ed. (1969). "Rein Control, or Unidirectional Rate Sensitivity, a Fundamental Dynamic and Organizing Function in Biology." *Ann. N. Y. Acad. Sci.* **156,** Art. 2, pp. 627–968.

Cunningham, W. J. (1958). "Introduction to Nonlinear Analysis." McGraw-Hill, New York.

DeVoe, R. D. (1962). Linear superposition of retinal action potentials to predict electrical flicker responses from the eye of the wolf spider. *J. Gen. Physiol.* **46,** 75–96.

DeVoe, R. D. (1963). Linear relations between stimulus amplitude and amplitudes of retinal action potentials from the eye of the wolf spider. *J. Gen. Physiol.* **47,** 13–32.

DeVoe, R. D. (1967). Nonlinear transient responses from light-adapted wolf spider eyes to changes in background illumination. *J. Gen. Physiol.* **50,** 1961–1991.

Fuortes, M. G. F. (1958). Electric activity of cells in the eye of limulus. *Amer. J. Ophthalmol.* **46,** 210–223.

Hardy, J. D., and Hammel, H. T. (1963). Control system in physiological temperature regulation, *in* "Temperature, Its Measurement and Control in Science and Industry," Vol. 3, pp. 613–625. Van Nostrand-Reinhold, Princeton, New Jersey.

Hensel, H. (1963). Electrophysiology of thermoreceptive nerve endings, *in* "Temperature, Its Measurement and Control in Science and Industry," Vol. 3, pp. 191–198. Van Nostrand-Reinhold, Princeton, New Jersey.

Nichols, J., Girling, F., Jerrard, W., Claxton, E. B., and Burton, A. C. (1951). Fundamental instability of the small blood vessels and critical closing pressure in vascular beds. *Amer. J. Physiol.* **164,** 330–344.

Urquhart, J., and Li, C. C. (1968). The dynamics of adrenocortical secretion. *Amer. J. Physiol.* **214,** 73–85.

Urquhart, J., Krall, R. L., and Li., C. C. (1968). Adrenocortical secretory function — Communications and control aspects. *IFAC Symp. Tech. Biol. Problems.* Erivan, Armenia, 1968.

Yamamoto, W. S. (1965). Homeostasis, continuity, and feedback, *in* "Physiological Controls and Regulations." Saunders, Philadelphia, Pennsylvania.

Biochemical Control ‖ *Chapter 10*

In the final analysis, the mechanisms by which all biological processes take place are chemical in nature and must be explained ultimately in the transformation of one chemical species into another, of the synthesis and catabolism of specific molecules. Many of these reactions take place within a cell, and most of them are catalyzed by appropriate enzymes. Some of these reactions can be described by linear differential equations, and when such is the case, the behavior of the reaction is very similar to that of the compartmental systems previously described.

On the other hand, probably the greater portion of biochemical reactions are nonlinear, leading to a variety of response patterns hardly suggested by the limited discussion of Chapter 9. Furthermore, the network of biochemical reactions within a single cell is controlled and coordinated by an array of control and feedback pathways, the diversity of which is far greater than the discussion in the previous chapters would indicate. It will be impossible to discuss biochemical control in any great detail, owing in part to space limitations and in part to our incomplete knowledge of the many mechanisms. Despite these limitations, we shall discuss some of the feedback systems that have been investigated, and attempt to place these results in a broader picture of physiological control.

One point of terminology needs clarification. We have previously used the concept of *order* to signify the order of the differential equation, or what is the same thing, the number of compartments in the system. This is accepted mathematical terminology. Unfortunately, in chemical kinetics, the term order has a quite different meaning. It refers to the number of molecules that combine to form a new one. In such a case, the rate of the reaction is proportional to the *product* of the concentrations of the two precursor molecules. A second-order chemical reaction leads to a nonlinear differential equation, containing products of dependent variables, and whose solution bears no similarity to those of the second-order equations previously discussed. To avoid ambiguity on this point we shall retain the term *order* to denote the mathematical order, and use *order of the chemical reaction,* or some similar phrase, when reference is made to the character of the chemical reaction.

This chapter describes certain rate processes for which the roles of stimulus and initial conditions are somewhat different than that encountered in previous chapters. The mathematical relations are developed from the conservation of mass, and the mass action law for chemical reactions. It will be noted that in most instances the final values are rates of change (production or destruction).

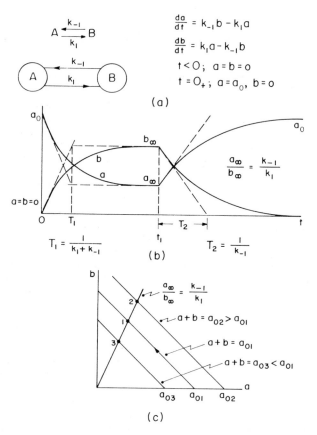

Fig. 10-1. Closed chemical system containing a reversible first-order reaction described by two first-order differential equations. (a) The stoichiometric and differential equations. (b) Response to a change in the concentration of A from 0 to a_0 at $t = 0$, and subsequently, at $t = t_1$, a reduction in k_1 to 0. (c) Phase plane portrait for this reaction with three different values for a_0, that is, a_{01}, a_{02}, a_{03}.

328

a. The chemical reaction represented by the equations and diagrams in this figure is a *closed system,* in that there are no sources or loss of material during the reaction. The quantities a and b signify the concentrations of the molecular species A and B. The reaction obeys the *law of mass action;* that is, the reaction rates are proportional to the concentrations of the reacting species (Higgins, 1965).

b. With both a and b initially zero, consider a sudden change in a from zero to a_0. The initial conditions are as shown below the equations, and the system then moves toward the steady state a_∞, b_∞. Although this system has two compartments, because it is closed, the response curves are single exponentials so that this is a first-order dynamical system. Note that the time constants, which may be readily derived from the differential equations, are different for the two transients.

c. The final steady-state values a_∞, b_∞, depend upon the total quantity of material present, but their ratio is determined solely by the rate constants as indicated in the figure.

d. A second disturbance occurs at t_1, at which instant the rate constant k_1 is presumed to be arbitrarily set equal to zero. Although such a change may not be readily realizable, it does provide further insight into process behavior. The concentrations return to their initial values, but with a longer time constant.

e. From a chemical standpoint, a disturbance may take either of the two forms described above. Changes in temperature, pressure, pH, and other factors can affect the k's, but with regard to biological processes it is probably the changes in concentration that are of prime, although not necessarily sole, significance.

f. The behavior of the system on the phase plane is shown at (c). Any point in the plane represents a specific pair of values of a and b. The lines with negative slope are for three values of a_0; the line of positive slope gives the ratio of a to b in the steady state. Their intersection yields the steady-state operating point. Although in the steady state $da/dt = db/dt = 0$, this does not imply that the forward and reverse reactions are both zero, but rather that they are equal.

g. Point 1 is the steady-state operating point with $a_0 = a_{01}$, whereas with $a_0 = a_{02}$, the final operating point is 2. The dynamic behavior is portrayed by the representative point as it moves along one of the trajectories from the initial to the final state. This motion is given by the solution to the differential equation as depicted in (b).

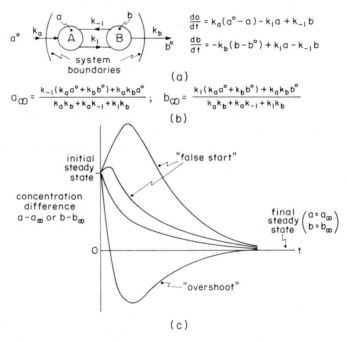

Fig. 10-2. Open system of two coupled first-order reactions yielding second-order dynamics. System boundaries distinguish between concentrations outside (a^o, b^o) from those inside (a, b) the system. The equations in (a) are derived from the law of mass action. The steady-state relations obtained by setting da/dt and db/dt equal to zero are given in (b). A variety of dynamic responses is possible, as shown at (c), but they all consist of only exponential terms (Denbigh, Hicks, and Page, 1948). Their form depends upon the relative magnitudes of the constants k_a, k_b, k_1, k_{-1}.

10   BIOCHEMICAL CONTROL

a. Biological processes are almost invariably open systems in which specified molecules move through the compartments, during which time chemical reactions take place. It is thus possible to identify input and output molecules which may be quite unlike, but which obey the usual conservation law for mass. The transformation may be associated with the release or storage of energy, or with the biosynthesis of specific molecules.

b. The ubiquity of open systems has led to their designation as *biochemical machines* (Green and Hatefi, 1961), the similarity rising from the fact that there is a continuous flow of energy or material through the system, and in the steady state the rate of production continues at a constant value.

c. The two-compartment example shown in (a) is termed a reversible first-order chemical reaction, but is described by two first-order differential equations and hence exhibits second-order dynamics. These equations are written on the basis that all flows obey the mass action law. However, it is quite possible that the flows between the environment and the compartments follow some other law.

d. The differential equations are in terms of the concentrations a and b. The steady state occurs when $da/dt = db/dt = 0$, and these relations then yield the values of a_∞ and b_∞ as given in the figure. Note that the two steady-state concentrations depend upon all six constants in the differential equations, but are independent of the initial concentrations.

e. The fact that the steady state is independent of the initial concentrations of a and b, that is, of the initial conditions, is an attribute of all linear systems. It is equivalent to saying that regardless of the initial state, the final values depend only upon the system parameters (the several rate constants) and the concentrations in the environment.

f. A change in any of the system parameters that produces a new steady state will result in any one of a number of dynamic behaviors as suggested in (c). It may be shown that only exponential terms can appear in the transient response, and therefore all the curves in (c) are composed of just two exponential terms. The variety of possible forms is brought about by differences in the two time constants together with the fact that each term may have a plus or minus sign. See Problem 3.

g. Reference to the discussion of dynamic behavior in Chapter 3 shows that the time constants depend upon the values of the four rate constants. On the other hand, the magnitude and sign of each exponential component in the transient depends upon the initial conditions.

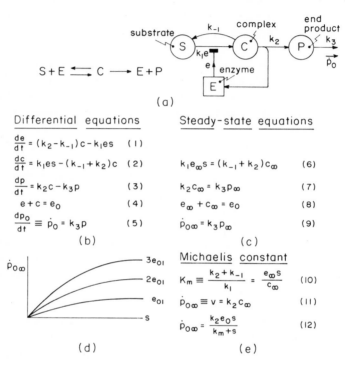

(a)

$$S + E \rightleftharpoons C \longrightarrow E + P$$

Differential equations

$$\frac{de}{dt} = (k_2 - k_{-1})c - k_1 es \quad (1)$$

$$\frac{dc}{dt} = k_1 es - (k_{-1} + k_2)c \quad (2)$$

$$\frac{dp}{dt} = k_2 c - k_3 p \quad (3)$$

$$e + c = e_0 \quad (4)$$

$$\frac{dp_0}{dt} \equiv \dot{p}_0 = k_3 p \quad (5)$$

(b)

Steady-state equations

$$k_1 e_\infty s = (k_{-1} + k_2)c_\infty \quad (6)$$

$$k_2 c_\infty = k_3 p_\infty \quad (7)$$

$$e_\infty + c_\infty = e_0 \quad (8)$$

$$\dot{p}_{0\infty} = k_3 p_\infty \quad (9)$$

(c)

(d)

Michaelis constant

$$K_m \equiv \frac{k_2 + k_{-1}}{k_1} = \frac{e_\infty s}{c_\infty} \quad (10)$$

$$\dot{p}_{0\infty} \equiv v = k_2 c_\infty \quad (11)$$

$$\dot{p}_{0\infty} = \frac{k_2 e_0 s}{k_m + s} \quad (12)$$

(e)

Fig. 10-3. The Michaelis–Menton equations for an enzyme-controlled reaction. (a) Stoichiometric equation and the flow diagram. (b) Differential equations for the several steps based upon the conservation law for each compartment. The variables s, c, p represent the number of molecules of each substance expressed in some suitable measure of concentration. e_0 is the number of active sites on the E molecules, of which e are unoccupied or free. Equation (4) is a conservation equation for active sites on E. (c) Steady-state equations. (d) Steady-state curves showing $p_{0\infty}$ to be a linear function of e and a saturating function of s. The subscript ∞ denotes the steady-state value. (e) Definition of the Michaelis constant K_m, and expressions for the steady-state reaction rate $p_{0\infty}$.

10   BIOCHEMICAL CONTROL

a.   Practically all biological reactions are associated with a specific enzyme (catalyst). Although the reactions will take place in the absence of the enzyme, only in its presence does the reaction proceed at a rate that has biological significance. In a sense, the enzyme can act as a control signal for the reaction, but the precise nature of the control may take many forms. The Michaelis–Menton equations shown in this figure serve as a prototype for most reactions catalyzed by an enzyme (Riggs, 1963).

b.   The reaction shown in (a) consists of the transformation of a substrate S into an end product P, under the influence of an enzyme E. A distinctive feature is the formation of an intermediate complex C, which probably represents the temporary binding of an S molecule onto an active site on the enzyme molecule.

c.   The "input" is considered to be the concentration s, and the "output" is the rate of production of P, namely $\dot{p}$. The rate at which P molecules are leaving the process is denoted by $\dot{p}_o$. The behavior of $\dot{p}$ depends upon the constraints placed upon the magnitude of e_o, the total number of active sites, and s, the available substrate. A common situation, and the one described here, is that in which e_o is fixed (Eq. 4), and s is maintained at a constant value, but one that may be altered by external conditions. Equations (1) and (2) are seen to be nonlinear owing to the product term es, but if s is maintained at a constant value the equations become linear.

d.   In the steady state, the differential equations reduce to Eqs. (6)–(9), from which Eqs. (10)–(12) are derived. The subscript ∞ has been used to emphasize that the relations apply only to the steady state. The Michaelis constant K_m becomes a matter of convenience in describing properties of the steady state.

e.   Part (d) of the figure is a plot of Eq. (12) in which the curves may be thought of as describing a control surface. For small values of s, the rate of production $\dot{p}_{o\infty}$ varies almost linearly with s, but this relation saturates for large values of s. Saturation arises because the binding sites tend to be more fully filled as s increases, and in the limit all binding sites are filled and the maximum value of $\dot{p}_{o\infty}$ is set by the *turnover number* for that enzyme; that is, the maximum number of molecules transformed per unit time. For each value of e_o, the maximum velocity $\dot{p}_{\infty\,max}$ occurs when all enzyme molecules are in the complex C.

f.   At saturation, the output $\dot{p}_\infty$ becomes only a function of e_o; the biochemical machine is operating at its maximum rate, as set by the enzyme concentration in the cell. Since the law of mass action is no longer applicable as regards s, this becomes a zero-order reaction.

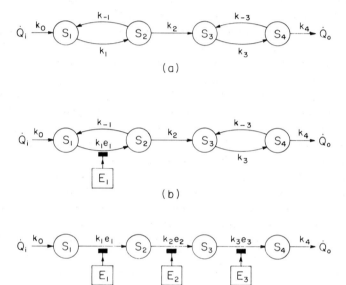

Fig. 10-4. Open systems in which a number of reactions occur sequentially. (a) Sequence of reversible and nonreversible reactions with the rate constants k_1, k_{-1}, etc. (b) Sequence in which one reaction is enzyme controlled, having the rate constant $k_1 e_1$. (c) Sequence of enzyme-controlled reactions with the rate constants $k_1 e_1$, etc.

10 BIOCHEMICAL CONTROL

a. The identification of one out of a number of processes operating sequentially, that in effect sets the pace for the entire system of reactions, is useful when considering the control aspects of a biosynthetic pathway. However, the concept of *rate-controlling*, or *rate-limiting*, has somewhat different meanings when applied to the steady state, on the one hand, and to dynamic behavior on the other (Bray and White, 1966).

b. Consider the sequence of processes in (a), where the reactions $S_1 \leftrightarrows S_2$, and $S_3 \leftrightarrows S_4$, are reversible, whereas that of $S_2 \rightarrow S_3$ is irreversible. Fluid flow models for these reactions are not always obvious; see Problem 5. In the steady state $\dot{Q}_i = \dot{Q}_o$, and the net flow between any two molecular species must also equal $\dot{Q}_i$. Conservation of mass requires that the flows equalize throughout the system, and it would be impossible to designate any one reaction as rate-controlling in the steady state, subject of course to the limitations that there is no saturation effect.

c. If one of the reactions is controlled by an enzyme as in (b), the conditions may be somewhat different. A fairly common situation is that in which the enzyme-catalyzed reaction operates with S_1 saturated so that the rate $k_1 s_1 e_1$ is dependent upon e_1 only and not upon s_1, as long as it is in a concentration well beyond saturation. The rate of conversion of S_1 to S_2, then, is not dependent upon the mass action law as regards S_1 and this catalyzed step is rate-limiting in the steady state.

d. An enzyme-controlled reaction may thus serve as the controlling reaction for an entire sequence of processes if the operation is in the region of substrate saturation. With a number of enzyme-controlled steps in sequence (c), the one with the smallest rate constant ke will be controlling. Furthermore, with a change in the system parameters, the control may be passed from one enzyme to another. If $k_k e_k$ is the rate-controlling step, no process following this one can have a greater rate.

e. With regard to the dynamic behavior of the entire system, the situation is quite similar to that described in earlier chapters. If the system were linear, then the process with the largest time constant would be the controlling one. All the processes with smaller time constants would approach a steady state in a relatively rapid manner, and the one with the largest time constant would in effect set the dynamic response for the whole system.

f. Zero-order processes may also be rate controlling. In an open system, no process subsequent to the one of zero order can have a greater reaction rate in the steady state. See Problem 7-10.

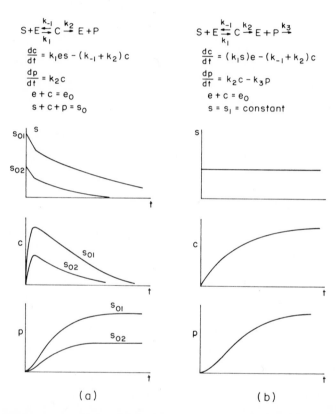

Fig. 10-5. Dynamic behavior of a single enzyme reaction as represented by the Michaelis–Menton equations. (a) Nonlinear behavior when total quantity of enzyme e_0 and substrate s_0 is fixed. (b) Linear behavior resulting when e_0 is fixed, and available substrate s is maintained constant. This implies that S is added to the reaction as fast as it is depleted. Curves are sketched from the paper by Chance (1960).

10   BIOCHEMICAL CONTROL

a. The previous sections have been largely concerned with the behavior of the enzyme-catalyzed reaction in the steady state, in which the reaction rates were all constant. This is termed the kinetic behavior of the process, and its characteristics will depend upon the various constraints placed upon the supply of E and S.

b. The dynamic behavior is that occurring when the system adjusts itself to a change in these constraints such as might be occasioned by a change in the total amount or effectiveness of E brought about by some control action, as discussed later. The time course followed by the various concentrations and reaction rates then becomes of interest.

c. Part (a) of the figure shows the set of equations governing the reaction when both S and E are present in fixed amount, s_o and e_o. This means that neither S nor E are added or lost during the ensuing reaction. When the reaction is initiated, all the substrate is present, but no P has been synthesized, and this figure depicts the course of events as S is depleted and P is formed. This is a closed system.

d. Inasmuch as the equations in (a) are nonlinear, as discussed previously, solutions to these and similar equations must be carried out with the aid of a computer. The relations and concepts developed earlier for linear systems can become very misleading if applied to nonlinear equations.

e. The behavior shown in (a) depicts C rising rapidly to a maximum value and then decreasing as P is synthesized. Although the location of the maximum in C and the precise shape of all the curves depend upon the relative values of the several rate constants, the ones selected here are believed to be reasonably representative. Curves are shown for different values of s_o.

f. In (b), it is assumed that P is removed as it is synthesized according to the mass action law $(-k_3p)$ and s is maintained at a constant value. The equations then become linear, and the system is described by two first-order linear differential equations. The response of C is a single exponential curve, and that of P is described by two exponentials. The assumption that S is maintained at a constant value is not too unrealistic if this process is but one step in a synthetic system for which there is an overall feedback control.

g. The temporal behaviors of intracellular reactions are, for the most part, fast compared to the physiological processes acting on the organism as a whole, and thus have negligible effect on many homeostats. However, an understanding of cellular processes will ultimately require an adequate picture of dynamics within the cell.

10.6 DYNAMICS OF THE MICHAELIS–MENTON EQUATIONS 337

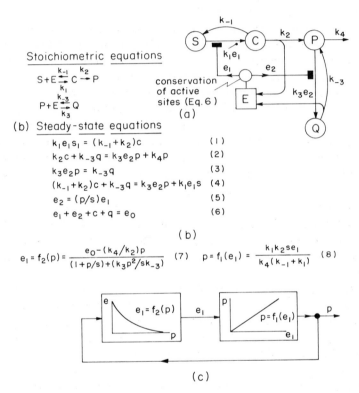

Stoichiometric equations

$$S + E \overset{k_2}{\underset{k_1}{\overset{k_{-1}}{\rightleftharpoons}}} C \rightarrow P$$

$$P + E \overset{k_{-3}}{\underset{k_3}{\rightleftharpoons}} Q$$

conservation
of active
sites (Eq. 6)

(a)

(b) Steady-state equations

$$k_1 e_1 s_1 = (k_{-1} + k_2) c \qquad (1)$$

$$k_2 c + k_{-3} q = k_3 e_2 p + k_4 p \qquad (2)$$

$$k_3 e_2 p = k_{-3} q \qquad (3)$$

$$(k_{-1} + k_2) c + k_{-3} q = k_3 e_2 p + k_1 e_1 s \qquad (4)$$

$$e_2 = (p/s) e_1 \qquad (5)$$

$$e_1 + e_2 + c + q = e_0 \qquad (6)$$

(b)

$$e_1 = f_2(p) = \frac{e_0 - (k_4/k_2) p}{(1 + p/s) + (k_3 p^2 / s k_{-3})} \qquad (7) \qquad p = f_1(e_1) = \frac{k_1 k_2 s e_1}{k_4 (k_{-1} + k_1)} \qquad (8)$$

(c)

Fig. 10-6. Control by competitive inhibition between S and P for a fixed number of active sites on E. The steady-state equations (1)–(6), when used to derive the transfer functions (7) and (8), yield the feedback system shown in (c).

a. Competition for the fixed number of active sites on E [e_o in Eq. (6)] occurs because E catalyzes both the reaction $S \rightarrow C$ and $P \rightarrow Q$. As p increases, more of the enzyme (e_2) is occupied in forming Q and that available for S, namely, e_1, is decreased.

b. The steady-state equations (1)–(6) are obtained from the differential equations in the same manner as in Fig. 10-3; they therefore represent the conservation principle for the several compartments. Note that Eqs. (2)–(4) contain nonlinear terms in the form of products of the dependent variables. Furthermore, Eq. (4) is the sum of Eqs. (1) and (3), and therefore is not an independent relation and may be disregarded.

c. The allocation of vacant binding sites ($e_1 + e_2$) to S and P is shown in Eq. (5) to be proportionate to the availability of these two molecules. This is believed to be a reasonable assumption.

d. The fact that the molecules of S and P compete for the same binding sites on E implies that they must have somewhat similar structures. Thus, competitive inhibition is restricted to those reactions in which this is true. Although this figure shows the competing molecule to be the end product P, other substrates could also serve this function.

e. In this figure, s is assumed to be a constant, so that in the steady state, the concentrations of all other molecules are also constant. There is thus a constant outflow $\dot{p}_o = k_4 p$.

f. The steady-state equations may be combined to yield the two transfer functions (7) and (8), which reveal the cyclic relations between the variables, and the negative slope of $f_2[p]$. Since the competition for enzyme assures a unilateral coupling, this system appears to satisfy all the operational rules of Section 4.3 required for a negative feedback system.

g. Examination of Eq. (6) shows that the quantity e_o serves the function of a reference input. Inasmuch as e_o is the total number of active sites on all the enzyme molecules, it will remain a constant as long as the number of E molecules is constant.

h. The sensitivity of this system to various kinds of disturbances may be investigated in the manner described in Sections 4.4 and 4.8. With a change in any of the system parameters, there will be a corresponding change in the characteristic curves in one or the other, or both, of the blocks. However, some of the rate coefficients appear in both transfer functions, so that the effect of feedback on sensitivity is by no means obvious.

i. This example illustrates one way in which competition between similar molecules serves to provide feedback control. Many other arrangements are possible.

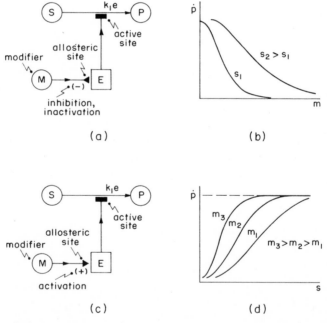

Fig. 10-7. Allosteric enzyme in which a modifier molecule bound to the enzyme molecule at an allosteric site can induce a conformational change at the active site of either sign. (a) Inhibition or inactivation by modifier reduces the reaction rate as shown in (b). (c) Modifier molecule can also enhance the reaction rate as shown in (d). Note the difference in axes used in (b) and (d) (Atkinson, 1965).

10   BIOCHEMICAL CONTROL

a. Allosteric enzymes are protein molecules having two or more kinds of active sites. In addition to the normal active site at which the substrate is bound, there are a fixed number of effector, or *modifier, sites* at which a small modifier molecule can be bound. The modifier (M) when bound to the enzyme alters the properties of the active or catalytic site so as to enhance or decrease the catalytic activity. This change within the enzyme molecule brought about by M is termed an *allosteric transformation,* and is considered to be a structural or conformational change within E (Monod, Wyman, and Changeux, 1965).

b. The diagram in (a) represents inhibition in which M has the ability to reduce or eliminate the catalytic activity of E. The modifier molecule M is reversibly bound to E and on the average a certain number of modifier sites will be occupied. The steady-state characteristic curves appear as in (b), where the rate of production of P, that is, $\dot{p}$, is plotted against concentration m. With m sufficiently large, all allosteric sites are occupied, and $\dot{p}$ falls nearly to zero.

c. Allosteric activation is also possible, as in the flow diagram shown in (c), which yields steady-state characteristics similar to those in (d). Activators and inhibitors have antagonistic effects as regards the active site, and both types of modifier molecules may be effective concurrently. The curves in (d) use s as the independent variable rather than m, but otherwise are quite similar to those in (b) except for the obvious difference in the slopes.

d. The characteristic curves for an allosteric enzyme are generally sigmoid in shape, rather than of hyperbolic form, as was the case in the simple enzyme reaction of Fig. 10-3. The sigmoid character suggests a higher-order reaction and some cooperative effect between a number of modifier molecules. The latter is suggested by the fact that $\dot{p}$ changes more rapidly than m at low values of m.

e. Inasmuch as the active and allosteric sites are separate regions on the enzyme molecule, there is no need for any structural similarity between the S and the M molecules. This was not the case in competitive inhibition (Fig. 10-6) where the two molecules P and S competed for the same active site. The allosteric enzyme thus opens up many possibilities for control because the modifier molecule can now control the reaction in which S takes part without the necessity of having any similarity of form. The allosteric enzyme thus acts as a transducer molecule in that one molecular species can control the reaction involving a quite different kind of molecule.

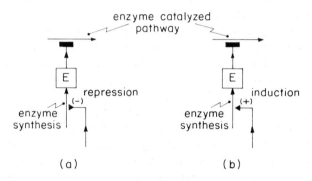

(a)   (b)

Fig. 10-8. This figure merely shows symbols employed to represent the repression and induction of enzyme synthesis, and is intended to be schematic only. Furthermore, these symbols represent the overall relations, and do not show the detailed steps by which genetic information is transcribed from the DNA molecule, nor do they show the roles played by the structural, operator, and regulator genes and the repressor molecule in the process of enzyme synthesis.

a. In engineering, and in particular in the development of technological systems, the designer is well advised to consider all conceivable possibilities regarding the components he selects, and the manner of their interconnection. Only by doing this can the engineer hope to approach an optimal solution. From the above standpoint, it is of interest to note that biochemical processes may have evolved in a similar manner, in that they appear to use all possible types of enzyme control, namely, control of enzyme effectiveness by activation and inhibition, and control of enzyme concentration in the manner described below.

b. The processes of *induction* and *repression* serve to enhance or retard the rate at which enzyme is synthesized within the cell and thus indirectly control the concentration. Inasmuch as enzyme molecules are continually being destroyed or lost to the cell by a variety of processes, these molecules must be continually synthesized at a rate necessary to maintain the normal complement.

c. Instructions for the synthesis of enzyme is contained within the DNA molecule, and this information is acted upon by RNA and repressor molecules. Although the precise molecular mechanisms by which the synthesis and repression is carried out are not known in great detail, a number of models have been proposed (Jacob and Monad, 1961). Despite the lack of detailed knowledge, there is ample experimental evidence to show that enzyme synthesis and repression are controllable.

d. The terms *induction* and *repression* are used to describe those processes that result in an increased or decreased concentration of enzyme within the cell. The symbols in the figure, though similar to the ones in Fig. 10-7 for activation and inhibition, are here applied to the input path for E. This is meant to signify a control of the number of enzyme molecules, whereas in the previous figure they implied a control of enzyme effectiveness (Umbarger, 1964).

e. Control of the rate at which enzyme molecules are synthesized becomes a usable control mechanism only because there is a normal turnover. Repression is thus a reduction in the rate of synthesis, so that the normal losses of enzyme will reduce the concentration of these molecules remaining in the cell. Induction is presumed to be a de-repression or interference with the repression mechanism.

f. Repression is a relatively slow process compared to inhibition of an allosteric molecule. The rates may differ by several orders of magnitude so that the dynamics of these two processes have quite different time scales. It has been suggested that, whereas activation and inhibition provide a fine control of the reaction, induction and repression are relatively coarse.

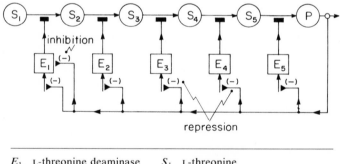

E_1	L-threonine deaminase	S_1	L-threonine
E_2		S_2	α-ketobutyrate
E_3	acetohydroxy acid isomeroreductase	S_3	α-acetohydroxybutyrate
		S_4	$\alpha\beta$-dihydroxy-β-methylvalerate
E_4	α-dehydrase	S_5	α-keto-β-methylvalerate
E_5	transaminase B	P	L-isoleucine

Fig. 10-9. Regulation of the L-isoleucine concentration (p) by two methods. Activity of the first enzyme E_1 is inhibited by end product P. The synthesis of all intermediate enzymes is repressed by P (Leavitt and Umbarger, 1961).

10 BIOCHEMICAL CONTROL

a. The previous sections have described a number of control mechanisms associated with the synthesis of specific molecules within a cell. The examples by no means exhaust the mechanisms that have been studied and proposed, but are sufficient, and sufficiently varied, to give some idea of the diversity of regulatory patterns that exist at the cellular level.

b. Examination of a specific synthetic pathway will usually reveal that more than one control mechanism is used, and that these several mechanisms are interconnected in a variety of ways. Some indication of this may be obtained from an examination of the pathway for L-isoleucine synthesis as shown in this figure (Changeux, 1965).

c. The pathway from threonine to L-isoleucine contains four intermediates, and the reactions yielding each of these species requires a specific enzyme, five enzymes in all. The overall feedback pathway is from the end product L-isoleucine to the first enzyme, threonine deaminase, whose activity is inhibited by binding of the isoleucine molecule.

d. A second mode of control is provided by the isoleucine molecules, which repress the synthesis of enzyme molecules. In a number of pathways, it has been found that repression acts on *all* intermediate enzymes so that with an excess of end product all intermediate enzymes are repressed. This arrangement, termed *coordination of enzyme repression,* is believed to be active in this synthetic pathway.

e. In feedback theory, coordinate repression would be likened to minor feedback loops from the regulated variable to each of the previous variables in the synthetic pathway. It would be logical to enquire as to the physiological significance of this arrangement as regards its effects upon the regulated variable. The latter is presumably the end product. Inasmuch as repression is a very slow process, these minor feedback loops can hardly have much to do with the dynamic behavior of the system.

f. A more reasonable explanation relates to the overall economy of material and energy. By repressing the synthesis of unneeded enzyme molecules, it should be possible to demonstrate an economy in the use of substrates and energy. However, further demonstration of the physiological implications of such economy is probably in order.

g. The induction of enzyme synthesis (not shown in the figure) is said to act by "turning on" a structural gene in the DNA molecule. Another possible mechanism is that of deactivating a repressor molecule, so-called de-repression.

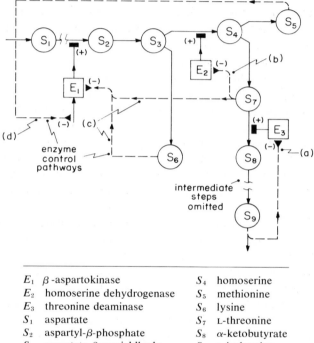

E_1	β-aspartokinase	S_4	homoserine
E_2	homoserine dehydrogenase	S_5	methionine
E_3	threonine deaminase	S_6	lysine
S_1	aspartate	S_7	L-threonine
S_2	aspartyl-β-phosphate	S_8	α-ketobutyrate
S_3	aspartate-β-semialdhyde	S_9	L-isoleucine

Fig. 10-10. Feedback control mechanisms in the synthesis of several amino acids. (a) Inhibition of an allosteric enzyme by feedback from an end product. (b) Feedback to an allosteric enzyme prior to a branch point leading to two different end products. (c) Concerted feedback from two products to a single enzyme. (d) Feedback from an end product producing enzyme repression (Stadtman, 1966).

a. The previous sections have described two modes of enzyme control: control of enzyme activity, and control of enzyme concentration. These by no means represent all possible variations, but they have been extensively studied, and as one examines synthetic pathways it becomes apparent that many combinations of these modes are possible.

b. Some of the pathways associated with the synthesis of four amino acids are shown in this figure, which has been simplified in the interests of clarity. At two points intermediate steps have been omitted, as indicated by the broken pathway. The intermediate steps leading to the synthesis of L-isoleucine have already been described in Fig. 10-9.

c. The feedback signals from one molecular species to an enzyme catalyzing some earlier step may take a number of forms. One common form is that of inhibition by an end product, sometimes termed end-product inhibition, in which a molecule exerts an inhibitory effect upon an enzyme appearing earlier in the synthetic pathway. This appears in (a), where E_3 controls the synthesis of S_8. A similar arrangement appears in (b): inhibition of E_2 activity regulates the synthesis of S_4 (Umbarger, 1961).

d. A second form of feedback is that of enzyme repression, shown in (d) as acting on E_1 in the pathway to S_2. Note that E_1 is an allosteric molecule that exerts control by both its transducer properties (conformational changes) as well as by repression of its synthesis.

e. A third form of feedback is shown in (c), where the joint feedback from two end products (S_6 and S_7) is directed to a single enzyme E_1. This is termed *concerted feedback,* because all end products must be present in excess amounts before there is inhibition of enzyme activity (Stadtman, 1966).

f. Another case is *cooperative feedback,* in which an excess of any of the end products causes a partial inhibition of the enzyme. In addition, the simultaneous excess of two or more end products results in greater inhibition than would be predicted from the sum of the individual effects. *Cumulative feedback* describes the case in which the inhibitory effects of the several end products act independently. It is also found that an enzyme at a certain step can take on a number of different forms, each inhibited by only one of the end products.

g. Although many allosteric enzymes control a reaction that is unique to one end product, note that E_3 catalyzes a step preceding a branch point. The consequences of such coupling between feedback loops is not clear.

Identification of the chemical processes and pathways making up a regulatory circuit poses the same kinds of problems as were encountered in other types of homeostats. Not only must the individual processes be identified, along with their input–output relations, but in addition the pathways from one process to another, together with all inputs and outputs, must be located. The identification problem associated with biochemical processes is undoubtedly more difficult than that encountered with the other homeostats, owing in large part to the problems of measurement.

The observation that the concentrations of certain chemical species are maintained at relatively constant values is at best only circumstantial evidence for the existence of a regulatory system. However, the application of the operational rules given on p. 89 should suffice for biochemical homeostats, although some extension of these criteria may well be necessary in the future as we gain more knowledge of the diversity of biochemical control processes. It has been pointed out by many individuals that the physiological importance of a postulated control mechanism must be demonstrated before it can be accepted as an essential relationship (Blakeley and Vitols, 1968).

The brief discussion in this chapter will have suggested that the objective of a regulatory process may not be simply that of maintaining some physical variable at a constant magnitude. Questions of economy have arisen and it may well be that other desiderata will be found that give rise to regulatory systems of hitherto unsuspected kinds.

If one considers the significance of homeostasis for the whole organism, it becomes clear that the homeostats related to temperature, blood, body chemistry and many other body processes have as their principle function that of maintaining the *milieu interieur* described by Claude Bernard. This is the very environment in which the cells live and breath, and in which the biochemical processes discussed in this chapter are immersed. It might be assumed, then, that the distrubances which upset the body homeostats are largely missing from the interior of the cell, but this is only partly true. In the first place, the conditions inside a single cell must be considered as a microenvironment, and only partially reflecting conditions in the body as a whole. In the second place, a very large number of similar cells are performing a given synthetic process, and there will undoubtedly be some reduction in the perturbations on a statistical basis alone.

In most instances, enzyme-controlled reactions operate in a region that is substrate saturated, and enzyme limited. That is, the reaction rate is set almost entirely by the properties and concentration of the catalyst, and is very little affected by changes in the concentration of the substrate molecules. Inasmuch as the substrate in a specific reaction is also the end product of a previous reaction, one would expect minimal changes in the substrate concentrations. However, a number of other factors such as temperature, pH, and the presence of small ions may have marked and pervasive effects upon the enzymes (Gutfreund, 1965).

Another aspect of feedback systems that should be mentioned in connection with biochemical processes is their propensity for oscillations and instability. The fact that biochemical systems can oscillate is now well known, and has been discussed in a number of publications (Higgins, 1967). The analysis of this problem is greatly complicated by the fact that the equations are nonlinear, and the simple methods of Chapter 9 are not adequate. Suffice it to say that chemical processes not greatly different from the ones described in this chapter can exhibit continuing oscillations, and thus may be unstable as a regulating system, or may become an oscillator and serve to generate some of the biological rhythms found quite generally in a wide variety of organisms.

Most of the experimental findings that underlie our knowledge of biochemical processes have been obtained from bacteria, among which there are considerable species variations involving the organization of presumably similar synthetic pathways. It is presumed, upon a minimum of evidence, that similar control mechanisms are present in mammalian cells.

The thoughtful reader may detect the fact that the organization of this book is in a sense the reverse of what evolutionary developments may have been. This text started with regulatory processes at the organ or organism level; quite possibly a consequence of the fact that homeostasis was first observed at these levels. The discussion eventually reached the cellular level, where it was found that the patterns of control equalled if not exceeded the number of varieties found at the higher levels of organization. Placing the two levels in juxtaposition, one senses many questions regarding the organization of regulating processes, and the possible effects of one level of organization upon the others as regards their homeostatic capabilities.

Epitome 10.13

1. This chapter describes a few feedback mechanisms at the biochemical level to emphasize the ubiquity of regulatory processes. Inasmuch as most biochemical reactions are of the second order chemically, they lead to nonlinear equations, which makes intuitive judgments regarding their behavior almost impossible.

2. In a closed system of first-order reactions, the final steady state depends upon the total material present and the several rate constants. A transient period will follow any change in these quantities.

3. Most biological processes are open systems in which there is a continuous flow of material or energy from an "input" to an "output," thus giving rise to the term "biochemical machines." In a second-order system having only exponential modes, a variety of transient behaviors is possible with the proper choice of initial conditions.

4. The simplest enzyme-catalyzed reaction, in which a single substrate forms a single intermediate complex, is represented by the Michaelis–Menton equations. This serves as a prototype for most enzyme controlled reactions; the intermediate complex representing the binding of a substrate molecule on an active site on the enzyme.

5. In a sequence of biochemical reactions, it is sometimes possible to identify one step as rate-controlling for the entire sequence. A catalyzed step is frequently rate-controlling, but one must distinguish between its dynamic and steady-state effects.

6. The Michaelis–Menton equations are inherently nonlinear owing to the presence of terms consisting of products of the substrates and enzyme concentrations. However, in a number of practical cases one or the other of these concentrations is constant, and the equations become linear.

7. Competition between substrate and end-product molecules for a limited number of active sites results in a negative feedback signal. The total number of active sites within the cell serves as a reference input for this feedback system.

8. Certain enzyme molecules are able to increase or decrease the effectiveness of their active sites as a result of the binding of another (modifier) molecule at an allosteric site. These allosteric enzymes can serve as "transducers" in that they control the binding of one molecular species by another quite different species.

9. Enzyme control is also accomplished by changing the number of enzyme molecules, and therefore the total number of active sites, within the cell. This is brought about by the induction and repression of enzyme synthesis.

10. The synthesis of L-isoleucine is controlled by allosteric inhibition of the first enzyme unique to its synthesis, and by coordinate repression of the synthesis of all enzymes active in this pathway. It appears that the coordinate repression serves the interest of overall cell economy regarding material and energy.

11. The inhibition of enzyme activity may result from signals from a number of end products, with these signals combined in various ways. Repression as well as inhibition may act upon a single enzyme. Although feedback signals from a given end product frequently act upon the first reaction that is unique to that molecule, this is not invariably the case, and feedback to reactions just preceding a branch point are found.

12. Although many of the disturbing factors that can affect biochemical reactions are minimized by other body homeostats, it is the microenvironment of the cell that ultimately determines its behavior. In addition to the usual reasons for requiring homeostasis, the economy of the cell provides another reason for needing feedback control.

Problems

1. The following questions refer to Fig. 10-1.

a. Prove that $a_\infty/b_\infty = k_{-1}/k_1$ as given in part (b) of the figure. Justify this relation on physical grounds.

b. Show that the differential equation describing each of the transients shown in (b) is of the first order. Calculate the time constant for each.

c. Is Fig. 10-1c drawn to the same scale as (b)? Redraw both (b) and (c) to show the case in which $a_\infty = b_\infty$.

d. Sketch the path on the phase plane if the ratio k_{-1}/k_1 is reduced by one half after reaching the operating point 1 in (c).

2. Derive the steady-state values a_∞, b_∞ as given in Fig. 10-2.

3. The response curves in Fig. 10-2c contain just two exponential components; the steady state is zero. In general, these components may be of either sign, and unequal in magnitude. With an initial state that is positive, show that only three cases are possible, and that these will yield response curves of the forms drawn in this figure.

4. Confirm the equations appearing in Fig. 10-3.

5. Sketch the tank, fluid-flow counterparts of the systems sketched in Fig. 10-4a, b, c. What might serve as a counterpart of the enzyme controlled reactions? See Problem 7-10.

References

Atkinson, D. E. (1965). Biological feedback control at the molecular level. *Science* **150**, 851–857.

Blakeley, R. L., and Vitols, E. (1968). The control of nucleotide biosynthesis. *Annu. Rev. Biochem.* **37**, 201–224.

Bray, H. G., and White, K. (1966). "Kinetics and Thermodynamics in Biochemistry," p. 208. Churchill, London.

Chance, B. (1960). Analogue and digital representations of enzyme kinetics. *J. Biol. Chem.* **235,** 2440–2443.

Changeux, J.-P. (1965). The control of biochemical reactions. *Sci. Amer.* **212** (Apr.), 36–45.

Denbigh, K. G., Hicks, M., and Page, F. M. (1948). The kinetics of open reaction systems. *Trans. Faraday Soc.* **44,** 479–494.

Green, D. E., and Hatefi, Y. (1961). The mitochondrian and biochemical machines. *Science* **133,** 13–18.

Gutfreund, H. (1965). "An Introduction to the Study of Enzymes." Wiley, New York.

Higgins, J. (1965). Dynamics and control in cellular reactions, *in* "Control of Energy Metabolism" (B. Chance *et al.,* eds.), pp. 13–46. Academic Press, New York.

Higgins, J. (1967). Theory of oscillating reactions. *Ind. Eng. Chem.* **59** (May), 18–62.

Jacob, F., and Monod, J. (1961). Genetic regulatory mechanism in the synthesis of proteins. *J. Mol. Biol.* **3,** 318–356.

Leavitt, R. I., and Umbarger, H. E. (1961). Isoleucine and valine metabolism in E. Coli. *J. Biol. Chem.* **236,** 2486–2491.

Monod, J., Wyman, J., and Changeux, J.-P. (1965). On the nature of allosteric transitions; a plausible model. *J. Mol. Biol.* **12,** 88–118.

Riggs, D. S. (1963). "The Mathematical Approach to Physiological Problems." Williams & Wilkins, Baltimore, Maryland.

Stadtman, E. R. (1966). Allosteric regulation of enzyme activity. *Advan. Enzymol.* **28,** 41–154.

Umbarger, H. E. (1961). Feedback control by endproduct inhibition. *Cold Spring Harbor Symp. Quant. Biol.* **26,** 301–312.

Umbarger, H. E. (1964). Intracellular regulatory mechanisms. *Science* **145,** 674–679.

Index

A

Acclimation, 280
Acclimatization, 280
ACS, *see* Adjustable control system
Adaptation, 209, 279–282
  physiological, 280
  to sinusoidal stimuli, 282
Adjustment, 279–281
Adjustable control system (ACS), 280
Allosteric, *see* Enzyme, allosteric
Amplitude of sinusoid, 167
Analog, 27
Angular velocity (ω), 147, 166, 167
Aqueous, *see* Eye
Astatic device, 113
Astatic system, 125
Attenuation curve, 177
Autoregulation, 121
  engineering usage, 283
  physiological usage, 284

B

Baroreceptor, properties of, 197
Baroreceptor reflex, *see* Blood pressure regulator
Bernard, C. (*milieu interieur*), 4, 5, 348
Block diagram, 31, 73
  nonuniqueness, 37
Blood pressure regulator
  circulatory system and, 121
  instability, 237
  properties of, 157
Bode plot, 177

C

Calcium regulation, 271
Cannon, W. B., 5
Capacity, 7
Cell, flow of CO_2, 9
Cerebellum, 67
Characteristic equation, 79, 141
  roots of, 79, 145

Characteristic steady-state curve, 29, 89
Chemical reaction
  closed, 329
  disturbances to, 329, 349
  enzyme controlled, 333
  open, 331
  order of, 327
  oscillatory, 349
  reversible, 329
  sensitivity of, 339
  steady-state, 328–333
  transients in, 331, 337
  zero-order, 333, 335
Cheyne–Stokes breathing, 239
Circulatory system, 8–11, 24, 32, 39, 110, 116, 121, 156, 237, 258–263, 270, 302
Clonus, 239
Compartment, 7, 9, 11
  loss function of, 13–15
  storage function of, 13, 49
  thermal, 10
Compensation, 121, 283
Complex number, 143
  algebra of, 205
Concentration gradient, 9, 273
Conservation of mass, 12, 13, 31
Constants of integration, 55, 59, 79
Continuous function, 135
Corner frequency, 175
Coupling
  asymmetrical, 23
  bilateral, 23
  unilateral, 21, 89
Cramer's rule, 37, 78
Cybernetics, 5

D

Damping envelop, 65
Density, liquid (δ), 12
Derivative, 53

Frequency spectrum (*cont.*)
  of ideal control mechanism, 181
  and interacting processes, 208
  line, 193
  linear approximation for, 176–179, 182–185
  linear coordinates in, 175
  logarithmic coordinates in, 177
  pass band, 201
  phase and amplitude relations for, 183, 209
  polar plot of, 237
  second-order, 179, 201
  third-order, 183
  time constant ratios and, 183

G

Gain, 29, 73
  closed-loop, 93
  critical, 218–221
  effect of on spectrum, 185
  open loop, 93, 157
    measurement of, 97, 103, 269
  small signal (incremental), 29
Glucose transport, 33

H

Hierarchical system, 285
Hippus, 238
Homeostasis, 3
  evolution of, 85
  history of, 4, 5
Homeostat, 3
Homeostatic index, 47
Homeostatic system(s)
  for amino acid synthesis, 347
  for blood pressure, 121, 237
  for calcium level, 271
  interactions between, 253
  for L-isoleucine synthesis, 345
  for neuromuscular control, 123
  power supply dynamics in, 278
  for respiration, 117, 263
  for temperature, 109, 250–263
  for TH level, 111
Homeotherm, 251
Hormones
  CT (calcitonin), 271
  PTH (parathormone), 271
  TH (thyroxine), 39, 111
  TRF (thyrotropin releasing factor), 111

TSH (thyrotropin), 39, 111
Hydaulic flow resistance, 12, 33, 117
Hypothalamus in temperature regulation, 257–261, 265
Hysteresis, 305

I

Implicit summing point, 97, 105
Impulse
  ideal, 75
  knee jerk, 67
  muscle twitch, 77
  transmission of, 63, 75
Incremental sensitivity, 100
Inhibition
  competitive, 339
  by end product, 345–347
  neural, 117, 122, 258, 362
Initial conditions, 55, 79, 320
  on the phase plane, 312–315
  for second-order system, 139
Input, 19
  generalized reference, 95
Instability
  elastic, 303
  exponential, 227
  oscillatory, 227
  thermal, 301
Integral, 49
Integration constants, 55, 64, 79

K

K, loop gain, 93
Knee jerk, 67

L

Laplace transform, 72–74, 140–142
Law of conservation of mass, 12, 13, 31
Law of mass action, 15, 328–333
Limit cycle, 314, 317
Line segment, 167
Linear system
  piece-wise, 299
  sinusoidal behavior of, 173
  superposition principle and, 47, 173
  test for, 173
Loop, open vs. closed, 93, 133
Loss function, 13–15
  control of, 109, 258, 259, 262, 263
  resistance parameter, 19
Low pass system, 208

M

Machine, biochemical, 331
Mass action law, 15, 329, 330, 333
Mathematical vs. physiological function, 13
Mayer waves, 237
Michaelis–Menton equation, 333, 337
Minimum phase system, 207
Mode of free vibration, 71
  change with feedback, 131, 142–145
  effect of gain, 149
  exponential mode, 143
  oscillatory mode, 147
    transmission through feedback system,
      151
Modulation, sinusoidal, 173
  pulse signals, 199
Multivariable control
  in engineering context, 285
  in physiological context, 285
Muscle
  agonist, antagonist, 35, 123, 261
  heat flow in, 11
  response to sinusoidal disturbance, 187,
    199
  response to step function, 77
  steady-state response in, 29, 77
  twitch response in, 77

N

Nerve, motor, 35, 199
Nonlinear equation, 337
Nonlinear system, 293
  chemical reaction represented by, 337
  effect of process sequence on, 309
  multiple operating points, 300–303
Nonlinear transducer, 29, 295, 299
Nonlinearity
  approximation, piecewise linear, 299
  asymmetry in response of, 295, 319
  saturation, 299
  threshold, 299
Nonminimum phase system, 207, 276
Norm, ensemble, temporal, 107, 119
Nyquist diagram, 237
Nystagmus, 223

O

Operating point, 88–91, 329
  changes in, 135, 267
  multiple, 301
  with and without feedback, 131

uniqueness of, 89
Optimal performance, 280
Order
  of chemical reaction, 327
  of differential equation, 78
  of diffusion process, 275
  of multicompartment system, 69
Oscillation
  angular velocity, 64
  damped, 64–67
  damping envelop, 65
  modes of free vibration, 71
Oscillator
  limit cycle, 305, 315
  linear vs. nonlinear, 242, 315
  relaxation, 317
  stability of, 315
Output, 19

P

Parameter, 19
  distrubuted, 273, 276
  lumped, 273
Parametric control, 311
Parametric space, 225, 233
Parathyroid gland, 271
Pass band, 201, 207, 208
Period of sinusoid, 167
Periodic function, 165
Perturbation, 91, 101
Phase angle (phase), 167
  component of spectrum, 175
Phasor, 167, 205
  algebra of, 205
  relations in oscillatory system, 203
Pituitary gland, 39
Plane
  complex, s, 142–147
  phase, 137, 312–315, 329
Poikilotherm, 251
Pole, 143
  complex, 142–147
  dominant, 143
  in s plane, 142–145
  real, 142–145
Process(es)
  diffusion, 9
  facultative, 311
  fluid flow, 13
  heat flow, 11
  interacting, noninteracting, 20–23
  order of, 55, 57

Process(es) (*cont.*)
  rate controlling, 335
  rate limiting, 335
  single compartment, 17
  zero order, 245, 335
Pulse
  train, 193
  transmission, 63, 75
Pupil reflex, 235

R

Radian, 167
Receptor
  baroreceptor, 197
  Golgi tendon organ, 123
  limulus eye, 319
  spindle, 123, 199
  temperature, 255, 319
  wolf spider eye, linear model, 297
Redundancy, 284
Reference input, 93, 105, 286
  generalized, 95
  implicit, 107
Reflex
  baroreceptor, 121, 157
  pupil, 235
  stretch, 123
Regulated quantities
  categories, 3, 4
  identification of, 257
Regulating system, *see also* Homeostatic
    system
  changes in the operating point, 267
  engineering design of, 249
  generalized block diagram for, 119, 265
  liquid level, 108
  objectives of, 348
  proportional, 113
  Type 0, Type 1, 113
Regulation
  as a mechanism, 3
  as a property, 2
Representative point, 137
Resonance
  appearance with closed loop, 201
  damping and, 209
  phase relations leading to, 203
  wolf spider eye and, 189
Respiratory system
  block diagram for, 262
  Cheyne–Stokes breathing and, 239

disturbances to, 27
oscillatory behavior of, 239
sign reversal in, 116
Response
  asymmetrical, 319
  forced oscillatory (sinusoidal), 171
  frequency, 171
  monotonic, 65
  nonlinear, 295
  sinusoidal, and adaptation, 209
  tests for linear, 297
Root of characteristic equation, 79, 142–145

S

Self-regulation, 283
Sensitivity, 100, 103
  technological regulator, 103
  unidirectional rate, 157, 319
Separatrix, 321
Servomechanism, 113
Set point, 118, 267
Shivering, 239, 261
Sign reversal, 89, 261
  by control of loss function, 263
  mechanism for, 117
  subtraction process in, 93
Signal, 40
  actuating, 106, 118
  manipulated variable, 118
  periodic, 191
  small, theory of, 29, 173, 207, 293, 297
Sinusoid (sine wave), 165–167
  magnitude of, 167, 175
Slope, 16, 53
Small signal analysis, *see* Signal, small,
    theory of
Space
  parametric, 225, 233
  phase, 321
Spectrum 175, *see also* Frequency spectrum
  of baroreceptor, 197
  of blood pressure regulator, 237
  magnitude of, 175
  of nerve-muscle preparation, 187
  of pupil reflex, 235
  of stretch reflex, 199
  of wolf spider eye, 189
Stability, 215, 240
  absolute, 241
  asymptotic, 241
  of blood pressure regulator, 237

Transmission (*cont.*)
  pulse, 63
  signal, 63
  sinusoidal, 168–175
  steady-state, *see* Gain
Transport, active, passive, 33
Transport lag, 229
  and stability, 231
  in circulatory system, 229
  in feedback system, 231
  in neural conduction, 229
  spectrum of, 229
  transfer function for, 229
Tremor
  muscular, 77
  neuromuscular, 155
Two-position control, 304–307
  differential gap, 305

U

Unbalance, 106

V

Variable, 19
  input, 19
  manipulated, 118, 121
  nonnegative, 285
  output, 19
  regulated, 87, 118
Vasomotor control, 117, 121, 157, 253, 263, 269, 303
Vibrations
  forced, 206, 223
  free, modes of, 71, 206, 313

Z

"Zero" frequency, 175, 208
Zero order, 245, 333, 335